6th International Conference on Computers, Management and Mathematical Sciences (ICCM 2020)

Held online due to COVID-19

Department of Electronics and Communication Engineering, North Eastern Regional Institute of Science and Technology (NERIST)

Arunachal Pradesh, India
23 November 2020

ISBN: 978-1-7138-2253-0

Printed from e-media with permission by:

Curran Associates, Inc.
57 Morehouse Lane
Red Hook, NY 12571

Some format issues inherent in the e-media version may also appear in this print version.

Printed with permission by Curran Associates, Inc. (2021)

For permission requests, please contact International Association of Academicians Connecting Scholars, Scientists and Engineers (IAASSE) at the address below.

International Association of Academicians Connecting Scholars, Scientists and Engineers (IAASSE)

www.iaasse.org
info@iaasse.org

Print ISBN: 978-1-7138-2253-0
eISBN: 978-1-7138-2254-7

Additional copies of this publication are available from:

Curran Associates, Inc.
57 Morehouse Lane
Red Hook, NY 12571 USA
Phone: 845-758-0400
Fax: 845-758-2633
Email: curran@proceedings.com
Web: www.proceedings.com

TABLE OF CONTENTS

GENERALIZED DIFFERENTIAL TRANSFORM METHOD FOR SOLUTION OF NONLINEAR HARMONIC OSCILLATOR EQUATION WITH FRACTIONAL ORDER DAMPING TERM

Deepanjan Das[1], Madan Mohan Dixit[2] and Chandra Prakash Pandey[3]
[1]Department of Mathematics, Ghani Khan Choudhury Institute of Engineering and Technology, Narayanpur, Malda,West Bengal-732141, India.
[2]Department of Mathematics, North Eastern Regional Institute of Science and Technology, Nirjuli, Itanagar, Arunachal Pradesh-791109, India.
[3]Department of Mathematics, North Eastern Regional Institute of Science and Technology, Nirjuli, Itanagar, Arunachal Pradesh-791109, India.

E-mail:gkcietdeepanjan@gmail.com

Abstract. In the present paper, Generalized Differential Transform Method (GDTM) is used for obtaining the approximate analytic solutions of nonlinear harmonic oscillator equation with fractional order damping term .The fractional derivatives are described in the Caputo sense.

Keywords: Fractional differential equations; Caputo fractional derivative; Generalized Differential Transform Method; Analytic solution.

Mathematical Subject Classification (2010): 26A33, 34A08, 35A22, 35R11, 35C10,74H10.

1. Introduction

Differential equations with fractional order are generalizations of classical differential equations of integer order and have recently been proved to be valuable tools in the modeling of many physical phenomena in various fields of science and engineering. By using fractional derivatives a lot of works have been done for a better description of considered material properties. Based on enhanced rheological models Mathematical modeling naturally leads to differential equations of fractional order and to the necessity of the formulation of the initial conditions to such equations. Recently, various analytical and numerical methods have been employed to solve linear and nonlinear fractional differential equations. The differential transform method (DTM) was proposed by Zhou [1] to solve linear and nonlinear initial value problems in electric circuit analysis. This method has been used for solving various types of equations by many authors [2-15]. DTM constructs an analytical solution in the form of a polynomial and different from the traditional higher order Taylor series method. For solving two-dimensional linear and nonlinear partial differential equations of fractional order DTM is further developed as Generalized Differential Transform Method (GDTM) by Momani, Odibat and Erturk in their papers [16-18].

Recently,Vedat Suat Ertiirka and Shaher Momanib applied generalized differential transform method to solve fractional integro-differential equations [19]. The GDTM is implemented to derive the solution of space-time fractional telegraph equation by Mridula Garg, Pratibha Manohar and Shyam L. Kalla[20]. Manish Kumar Bansal and Rashmi Jain applied generalized differential transform method to solve fractional order Riccati differential equation [21]. Aysegul Cetinkaya, Onur Kiymaz and Jale Camli applied generalized differential transform method to solve nonlinear PDE's of fractional order [22].

2. Mathematical Preliminaries on Fractional Calculus

In the present analysis we introduce the following definitions [23, 24].

2.1. Definition: Let $\alpha \in R^{+}$ .On the usual Lebesgue space $L(a,b)$ integral operator I^{α} defined by

$$I^{\alpha} f(x) = \frac{d^{-\alpha} f(x)}{dx^{-\alpha}} = \frac{1}{\Gamma(\alpha)} \int_{0}^{x} (x-t)^{\alpha-1} f(t)\, dt$$

and

$$I^0 f(x) = f(x)$$

is called Riemann-Liouville fractional integral operator of order $\alpha \geq 0$ and $a \leq x < b$.
It has the following properties:

I. $I^\alpha f(x)$ exists for any $x \in [a,b]$,

II. $I^\alpha I^\beta f(x) = I^{\alpha+\beta} f(x)$,

III. $I^\alpha I^\beta f(x) = I^\beta I^\alpha f(x)$,

IV. $I^\alpha x^\gamma = \dfrac{\Gamma(\gamma+1)}{\Gamma(\alpha+\gamma+1)} x^{\alpha+\gamma}$,

where $f(x) \in L[a,b]$, $\alpha, \beta \geq 0$, $\gamma > -1$.

2.2. Definition: The Riemann-Liouville definition of fractional order derivative is

$$ {}_0^{RL}D_x^\alpha f(x) = \frac{d^n}{dx^n} {}_0I_x^{n-\alpha} f(x) = \frac{1}{\Gamma(n-\alpha)} \frac{d^n}{dx^n} \int_0^x (x-t)^{n-\alpha-1} f(t)\,dt,$$

where n is an integer that satisfies $n-1 < \alpha < n$.

2.3. Definition: A modified fractional differential operator ${}_0^cD_x^\alpha$ proposed by Caputo is given by

$$ {}_0^cD_x^\alpha f(x) = {}_0I_x^{n-\alpha} \frac{d^n}{dx^n} f(x) = \frac{1}{\Gamma(n-\alpha)} \int_0^x (x-t)^{n-\alpha-1} f^{(n)}(t)\,dt,$$

where $\alpha\left(\alpha \in R^+\right)$ is the order of operation and n is an integer that satisfies $n-1 < \alpha < n$.
It has the following two basic properties [25]:

I. If $f \in L_\infty(a,b)$ or $f \in C[a,b]$ and $\alpha > 0$, then ${}_0^cD_x^\alpha {}_0I_x^\alpha f(x) = f(x)$.

II. If $f \in C^n[a,b]$ and if $\alpha > 0$, then ${}_0I_x^\alpha {}_0^cD_x^\alpha f(x) = f(x) - \sum_{k=0}^{n-1} \dfrac{f^{(k)}(0^+)}{k!} x^k$; $n-1 < \alpha < n$.

2.4. Definition: For m being the smallest integer that exceeds α, the Caputo time-fractional derivative operator of order $\alpha > 0$, is defined as[26]

$$D_t^\alpha u(x,t) = \frac{\partial^\alpha u(x,t)}{\partial t^\alpha} = \begin{cases} \dfrac{\partial^m u(x,\xi)}{\partial \xi^m} & ; \quad \alpha = m \in N \\ \dfrac{1}{\Gamma(m-\alpha)} \displaystyle\int_0^t (t-\xi)^{m-\alpha-1} \frac{\partial^m u(x,\xi)}{\partial \xi^m} d\xi & ; \quad m-1 \leq \alpha < m \end{cases}.$$

Relation between Caputo derivative and Riemann-Liouville derivative:

$$ {}_0^cD_x^\alpha f(x) = {}_0^{RL}D_t^\alpha f(x) - \sum_{k=0}^{m-1} \frac{f^{(k)}(0^+)}{\Gamma(k-\alpha+1)} x^{k-\alpha} ; m-1 < \alpha < m.$$

Integrating by parts, we get the following formulae as given by [27]

I. $\displaystyle\int_a^b g(x)\, {}_a^cD_x^\alpha f(x)\,dx = \int_a^b f(x)\, {}_x^{RL}D_b^\alpha g(x)\,dx + \sum_{j=0}^{n-1} \left[{}_x^{RL}D_b^{\alpha+j-n} g(x)\, {}_x^{RL}D_b^{n-j-1} f(x) \right]_a^b$.

II. For $n = 1$, $\displaystyle\int_a^b g(x)\, {}_a^cD_x^\alpha f(x)\,dx = \int_a^b f(x)\, {}_x^{RL}D_b^\alpha g(x)\,dx + \left[{}_xI_b^{1-\alpha} g(x).f(x) \right]_a^b$.

3. Generalized one dimensional differential transform method

Generalized differential transform of a function $x(t)$ in one variable is denoted by $X_\alpha(k)$ and defined as follows[16-18]:

$$X_\alpha(k)=\frac{1}{\Gamma(\alpha k+1)}\left[\left(D_{t_0}^{\alpha}\right)^k x(t)\right]_{t=t_0}, \tag{1}$$

where $\alpha\in(0,1]$ and $\left(D_{t_0}^{\alpha}\right)^k = D_{t_0}^{\alpha}, D_{t_0}^{\alpha}, \ldots, D_{t_0}^{\alpha}\,(k-\text{times})$,

and the inverse generalized differential transform of $X_\alpha(k)$ is given by

$$x(t)=\sum_{k=0}^{\infty}X_\alpha(k)(t-t_0)^{\alpha k} . \tag{2}$$

It has the following properties:

I. If $x(t)=v(t)\pm w(t)$, then $X_\alpha(k)=V_\alpha(k)\pm W_\alpha(k)$.

II. If $x(t)=av(t);\ a\in R$, then $X_\alpha(k)=aV_\alpha(k)$.

III. If $X(t)=v(t)w(t)$, then $X_\alpha(k)=\sum_{r=0}^{k}V_\alpha(r)W_\alpha(k-r)$.

IV. If $x(t)=(t-t_0)^{n\alpha}$, then $X_\alpha(k)=\delta(k-n)$.

V. If $x(t)=D_{t_0}^{\alpha}v(t);\ 0<\alpha\le 1$, then $X_\alpha(k)=\frac{\Gamma(\alpha(k+1)+1)}{\Gamma(\alpha k+1)}V_\alpha(k+1)$.

VI. If $x(t)=t^{\lambda}f(t)$, where $\lambda>-1$, $f(t)$ has the generalized Taylor series expansion $f(t)=\sum_{n=0}^{\infty}a_n(t-t_0)^{n\alpha}$ with

   a. $\beta<\lambda+1$ and α is arbitrary or

   b. $\beta\ge\lambda+1$, α is arbitrary and $a_n=0$ for $n=0,1,2,\ldots m-1$, with $m-1<\beta\le m$. Then (1) becomes

$$X_\alpha(k)=\frac{1}{\Gamma(\alpha k+1)}\left[D_{t_0}^{\alpha k}x(t)\right]_{t_0}.$$

VII. If $x(t)=D_{t_0}^{\gamma}f(t)$, $m-1<\gamma\le m$ and the function $f(t)$ satisfies the conditions given in (VI), then $X_\alpha(k)=\frac{\Gamma(\alpha k+\gamma+1)}{\Gamma(\alpha k+1)}F_\alpha\left(k+\frac{\gamma}{\alpha}\right)$.

Where $U_\alpha(k), V_\alpha(k), W_\alpha(k)$ and $F_\alpha(k)$ are the differential transformations of the functions $u(t), v(t), w(t)$ and $f(t)$ respectively and

$$\delta(k-n)=\begin{cases}1 & ;\ k=n\\ 0 & ;\ k\ne n\end{cases}.$$

4. Test Problems

In this section, we present three examples to illustrate the applicability of Generalized Differential

Transform Method (GDTM) to solve nonlinear differential equations of fractional order.

4.1 Example: Consider the nonlinear fractional differential equation

$$\frac{d^2x(t)}{dt^2}+\omega^2x^2(t)+2k\frac{d^{1/2}x(t)}{dt^{1/2}}=0,$$

subject to initial conditions $x(0)=p$ (constant) and $x'(0)=q$ (constant) (3)

where $\frac{d^{1/2}}{dt^{1/2}}$ is the fractional differential operator(Caputo derivative) and k,ω are the damping coefficient and frequency of the oscillation respectively.

Applying generalized one-dimensional differential transform (1) with $t_0=0$ on (3), we obtain

$$X_{1/2}(h)=-\frac{\Gamma\left(\frac{1}{2}(h-4)+1\right)}{\Gamma\left(\frac{1}{2}(h-4)+3\right)}\left\{\omega^2\sum_{l=0}^{h-4}X_{1/2}(l)X_{1/2}(h-4-l)+2k\frac{\Gamma\left(\frac{1}{2}(h-4)+\frac{3}{2}\right)}{\Gamma\left(\frac{1}{2}(h-4)+1\right)}X_{1/2}(h-3)\right\}, \quad (4)$$

with $X_{1/2}(0)=p$ and $X_{1/2}(2)=q$. (5)

Now utilizing the recurrence relation (4) and the initial condition (5), we obtain after a little simplification the following values of $X_{1/2}(k)$ for $k=0,1,2,\ldots\ldots$

$$X_{1/2}(1)=0;\ X_{1/2}(3)=-\frac{2}{\Gamma\left(\frac{5}{2}\right)}kp;\ X_{1/2}(4)=-\frac{1}{2}\omega^2p^2;\ X_{1/2}(5)=-\frac{2}{\Gamma\left(\frac{7}{2}\right)}kq;$$

$$X_{1/2}(6)=-\frac{1}{3}p\left(\omega^2q-2k^2\right);X_{1/2}(7)=\frac{6}{\Gamma\left(\frac{11}{2}\right)}k\omega^2p^2;\ X_{1/2}(8)=\frac{2}{\Gamma(5)}\left\{\omega^2\left(\omega^2p^3-q^2\right)+k^2q\right\};$$

$$X_{1/2}(9)=\frac{4}{\Gamma\left(\frac{11}{2}\right)}\left(\frac{9}{2}\omega^2q-2k^2\right)$$

and so on.

Using the above values of $X_{1/2}(k)$; $k=0,1,2,\ldots\ldots$in (2), the solution of (3)is obtained as

$$x(t)=p+qt-\frac{2}{\Gamma\left(\frac{5}{2}\right)}kpt^{3/2}-\frac{1}{2}\omega^2p^2t^2-\frac{2}{\Gamma\left(\frac{7}{2}\right)}kqt^{5/2}-\frac{1}{3}p\left(\omega^2q-2k^2\right)t^3+\frac{6}{\Gamma\left(\frac{11}{2}\right)}k\omega^2p^2t^{7/2}$$
$$+\frac{2}{\Gamma(5)}\left(\omega^2\left(\omega^2p^3-q^2\right)+k^2q\right)t^4+\frac{4}{\Gamma\left(\frac{11}{2}\right)}\left(\frac{9}{2}\omega^2q-2k^2\right)t^{9/2}+\ldots\ldots\ldots\ldots\ldots\ldots \quad (6)$$

4.2 Example: Consider the nonlinear fractional differential equation

$$\frac{d^2x(t)}{dt^2}-\omega^2x^2(t)+2kx(t)\frac{d^{1/2}x(t)}{dt^{1/2}}=0,$$

subject to initial conditions $x(0)=p$ (constant) and $x'(0)=q$ (constant). (7)

where $\frac{d^{1/2}}{dt^{1/2}}$ is the fractional differential operator(Caputo derivative) and k,ω are the damping coefficient and frequency of the oscillation respectively .

Applying generalized one-dimensional differential transform (1) with $t_0=0$ on (7), we obtain

$$X_{1/2}(h)=\frac{\Gamma\left(\frac{1}{2}(h-4)+1\right)}{\Gamma\left(\frac{1}{2}(h-4)+3\right)}\left\{\omega^2X_{1/2}(h-4)-2k\sum_{l=0}^{h-4}X_{1/2}(l)\frac{\Gamma\left(\frac{1}{2}(h-l-4)+\frac{3}{2}\right)}{\Gamma\left(\frac{1}{2}(h-l-4)+1\right)}X_{1/2}(h-l-3)\right\}, \tag{8}$$

with $X_{1/2}(0)=p$ and $X_{1/2}(2)=q$. (9)

Now utilizing the recurrence relation (8) and the initial condition (9), we obtain after a little simplification the following values of $X_{1/2}(k)$ for $k=0,1,2,......$

$$X_{1/2}(1)=0;X_{1/2}(3)=0;X_{1/2}(4)=\frac{1}{2}\omega^2p;X_{1/2}(5)=-\frac{2}{\Gamma\left(\frac{7}{2}\right)}kpq;X_{1/2}(6)=\frac{1}{6}\omega^2q;$$

$$X_{1/2}(7)=-\frac{3}{\Gamma\left(\frac{9}{2}\right)}k\left(\frac{2}{3}p^2\omega^2+q^2\right);X_{1/2}(8)=\frac{1}{12}p\left(\frac{1}{2}\omega^4+2k^2pq\right);X_{1/2}(9)=-\frac{43}{4\Gamma\left(\frac{11}{2}\right)}\omega^2kpq$$

Using the above values of $X_{1/2}(k);\ k=0,1,2,.....$in (2),the solution of (7) is obtained as

$$x(t)=p+qt+\frac{1}{2}\omega^2pt^2-\frac{2}{\Gamma\left(\frac{7}{2}\right)}kpqt^{5/2}+\frac{1}{6}\omega^2qt^3-\frac{3}{\Gamma\left(\frac{9}{2}\right)}k\left(\frac{2}{3}p^2\omega^2+q^2\right)t^{7/2}$$
$$+\frac{1}{12}p\left(\frac{1}{2}\omega^4+2k^2pq\right)t^4-\frac{43}{4\Gamma\left(\frac{11}{2}\right)}\omega^2kpqt^{9/2}+............... \tag{10}$$

4.3 Example: Consider the nonlinear fractional differential equation

$$\frac{d^2x(t)}{dt^2}+\omega^2x^2(t)-2kx(t)\frac{d^{1/2}x(t)}{dt^{1/2}}=0,$$

subject to initial conditions $x(0)=p$ (constant) and $x'(0)=q$ (constant), (11)

where $\frac{d^{1/2}}{dt^{1/2}}$ is the fractional differential operator(Caputo derivative) and k,ω are the damping coefficient and frequency of the oscillation respectively .

Applying generalized one-dimensional differential transform (1) with $t_0=0$ on (11), we obtain

$$X_{\frac{1}{2}}(h)=\frac{\Gamma\left(\frac{1}{2}(h-4)+1\right)}{\Gamma\left(\frac{1}{2}(h-4)+3\right)}\left\{2k\sum_{l=0}^{h-4}X_{\frac{1}{2}}(l)\frac{\Gamma\left(\frac{1}{2}(h-l-4)+\frac{3}{2}\right)}{\Gamma\left(\frac{1}{2}(h-l-4)+1\right)}X_{\frac{1}{2}}(h-l-3)-\omega^2\sum_{l=0}^{h-4}X_{\frac{1}{2}}(l)X_{\frac{1}{2}}(h-l-4)\right\}, \tag{12}$$

with $X_{\frac{1}{2}}(0)=p$ and $X_{\frac{1}{2}}(2)=q$. (13)

Now utilizing the recurrence relation (12) and the initial condition (13), we obtain after a little simplification the following values of $X_{\frac{1}{2}}(k)$ for $k=0,1,2,......$

$$X_{\frac{1}{2}}(1)=0;X_{\frac{1}{2}}(3)=0;X_{\frac{1}{2}}(4)=-\frac{1}{2}\omega^2p^2;X_{\frac{1}{2}}(5)=\frac{2}{\Gamma\left(\frac{7}{2}\right)}kpq;X_{\frac{1}{2}}(6)=-\frac{1}{6}\omega^2p^2q^2;$$

$$X_{\frac{1}{2}}(7)=\frac{2}{\Gamma\left(\frac{9}{2}\right)}k\left(\frac{5}{2}q^2-\omega^2p^3\right);X_{\frac{1}{2}}(8)=\frac{1}{12}\left(2k^2p^2q-\omega^2\left(\frac{1}{4}\omega^4p^6+q^2\right)\right);$$

$$X_{\frac{1}{2}}(9)=-\frac{2}{\Gamma\left(\frac{11}{2}\right)}k\omega^2p^2q\left(pq+\frac{2}{\Gamma\left(\frac{7}{2}\right)}kp^2q+\frac{91}{40}kp^2q\right)\text{and so on}$$

Using the above values of $X_{\frac{1}{2}}(k)$; $k=0,1,2,.....$in (2), the solution of (11) is obtained as

$$x(t)=p+qx-\frac{1}{2}\omega^2p^2t^2+\frac{2}{\Gamma\left(\frac{7}{2}\right)}kpqt^{\frac{5}{2}}-\frac{1}{6}\omega^2p^2q^2t^3+\frac{2}{\Gamma\left(\frac{9}{2}\right)}k\left(\frac{5}{2}q^2-\omega^2p^3\right)t^{\frac{7}{2}}$$

$$+\frac{1}{12}\left(2k^2p^2q-\omega^2\left(\frac{1}{4}\omega^4p^6+q^2\right)\right)t^4-\frac{2}{\Gamma\left(\frac{11}{2}\right)}k\omega^2p^2q\left(pq+\frac{2}{\Gamma\left(\frac{7}{2}\right)}kp^2q+\frac{91}{40}kp^2q\right)t^{\frac{9}{2}}+...................... \tag{14}$$

5. Conclusion

In the present study, we have applied the Generalized Differential Transform Method (GDTM) to find the approximate analytic solutions of three examples of nonlinear harmonic oscillator equations with fractional order damping term. It may be concluded that GDTM is a reliable technique to handle linear and nonlinear fractional differential equations.GDTM provides more realistic series solutions compared with other approximate methods.

References

[1] Zhou, J.K.(1986).Differential Transformation and Its Applications for Electrical Circuits. Huazhong University Press, Wuhan, China.

[2] Chen, C.K., and Ho, S.H.(1999).Solving partial differential equations by two dimensional differential transform method. *Appl. Math. Comput.*, 106:171–179.

[3] Ayaz, F.(2004).Solutions of the systems of differential equations by differential transform Method. *Appl. Math. Comput.*, 147:547–567.

[4] Abazari, R., and Borhanifar, A.(2010).Numerical study of the solution of the Burgers and coupled Burgers equations by a differential transformation method. *Comput. Math. Appl.*, 59:2711–2722.

[5] Chen, C.K.(1999).Solving partial differantial equations by two dimensional differential transformation

Method. *Appl. Math. Comput.*, 106:171–179.
[6] Jang, M.J., and Chen,C.K.(2001).Two-dimensional differential transformation method for partial differantial equations. *Appl. Math. Comput.*, 121:261–270.
[7] Kangalgil, F., and Ayaz, F.(2009).Solitary wave solutions for the KDV and mKDV equations by differential transformation method.*Choas Solitons Fractals*,41:464–472.
[8] Arikoglu, A., and Ozkol, I.(2006).Solution of difference equations by using differential transformation Method. *Appl. Math. Comput.*, 174:1216–1228.
[9] Momani,S., Odibat, Z., and Hashim,I.(2008).Algorithms for nonlinear fractional partial differantial equations: A selection of numerical methods. *Topol. Method Nonlinear Anal.*,31:211–226.
[10] Arikoglu,A., and Ozkol,I.(2007).Solution of fractional differential equations by using differential transformation method.*Chaos Solitons Fractals*, 34:1473–1481.
[11] Soltanalizadeh,B.,and Zarebnia,M.(2011).Numerical analysis of the linear and nonlinear Kuramoto-Sivashinsky equation by using differential transformation method. *Inter. J. Appl. Math. Mechanics*,7(12):63–72.
[12] Tari,A.,Rahimi,M.Y.,Shahmoradb,S.,and Talati,F.(2009).Solving a class of two-dimensional linear and nonlinear Volterra integral equations by the differential transform method.*J. Comput. Appl. Math.*, 228:70–76.
[13] Nazari,D., and Shahmorad,S.(2010).Application of the fractional differential transform method to fractional-order integro-differential equations with nonlocal boundary conditions.*J. Comput. Appl. Math.*, 234:883–891.
[14] Borhanifar,A., and Abazari,R.(2011).Exact solutions for non-linear Schr.dinger equations by differential transformation method.*J. Appl. Math. Comput.*,35:37–51.
[15] Borhanifar,A., and Abazari,R.(2010).Numerical study of nonlinear Schr.dinger and coupled Schr. dinger equations by differential transformation method.*Optics Communications*,283:2026–2031.
[16] Momani,S.,Odibat,Z.,and Erturk,V.S.(2007).Generalized differential transform method for solving a space- and time-fractional diffusion-wave equation.*Physics Letters. A*,370(5-6):379–387.
[17] Odibat,Z., and Momani,S.(2008).A generalized differential transform method for linear partial differential equations of fractional order. *Applied Mathematics Letters*, 21(2):194–199.
[18] Odibat,Z.,Momani,S.,and Erturk,V.S.(2008).Generalized differential transform method:application to differential equations of fractional order. *Applied Mathematics and Computation*, 197(2):467–477.
[19] Ertiirka,V.S., and Momanib,S.(2010).On the generalized differential transform method: application to fractionalintegro-differential equations. *Studies in Nonlinear Sciences*,1(4):118-126.
[20] Garg,M.,Manohar,P., and Kalla,S.L.(2011).Generalized differential transform method to Space-time fractional telegraph equation. *Int.J.of Differential Equations,*Hindawi Publishing Corporation, vol.2011,article id.:548982,9 pages,doi.:10.1155/2011/548982.
[21] Bansal,M.K., and Jain,R.(2015).Application of generalized differential transform method to fractional order Riccati differential equation and numerical results. *Int. J.of Pure and Appl. Math.*,99(3):355-366.
[22] Cetinkaya,A.,Kiymaz,O., and Camli,J.(2011).Solution of non linear PDE's of fractional order with generalized differential transform method. *Int. Mathematical Forum*,6(1):39-47.
[23] Das.S.(2008).Functional Fractional Calculus. Springer.
[24] Miller,K.S., and Ross,B.(1993).An Introduction to the Fractional Calculus and Fractional Diff. Equations. John Wiley and Son.
[25] Caputo,M.(1967).Linear models of dissipation whose q is almost frequency independent-ii. *Geophys J. R. Astron. Soc*,13:529-539.
[26] Podlubny,L.(1999).Fractional differential equations: An introduction to fractional derivatives, fractional differential equations, to methods of their solution and some of their applications. Academic Press.
[27] Almeida,R., and Torres,D.F.(2011).Necessary and sufficient conditions for the fractional calculus of variations with caputo derivatives. *Communications in Nonlinear Science and Numerical Simulation*, 16:1490-1500.

Effective Sprinkler irrigation system for tea cultivation: WSN based design

Hemarjit Ningombam[1], Dr. Om Prakash Roy[2]

[1]Dibrugarh University, Dibrugarh , Assam.
[2]North Eastern Regional Institute of Science and Technology (NERIST) Nirjuli, Arunachal Pradesh

e-mail:nhsingh@dibru.ac.in

Abstract. This paper describes the design of a simple and effective way of implementing sprinkler systems for irrigation in the tea field, utilizing a mesh capable wireless sensor network for monitoring and controlling field irrigation systems. Irrigation systems will effectively acquire information related to the quality and quantity of parameters and specific area of coverage required from the distributed sensors nodes. Accordingly, the feedback control will be executed to the specified sprinkler(s) system associated with the information acquired. Embedded multi-sensing and actuator nodes are distributed around the farms and used to measure the parameters necessary in controlling the irrigation process. With their positions information, farm-level information is transmitted to the ZIGBEE wireless sensor networks' centralized management system. Accordingly, the necessary control signal will generate and send only at the targeted positions or area to control feedback action on time. In this method, water is forced through nozzles in the form of a jet over the air and ensured to fall on the crop field resembling rainfall. The correct pressure of water application is provided through monitoring of the nozzle pressure using a pressure gauge. It has become necessary to develop an effective irrigation scheduling system so that the efficiency of irrigation systems and the prevention of water waste can be improved. The process will enable the farmers to optimize irrigation, pesticides, fertilizers, etc. only when and where it is needed.

I. INTRODUCTION

The Generally in large area cultivation, normal agricultural processes and standard farm management techniques were practices irrespective of real time conditions of the field. Fertilizers, insecticides, waters etc. requirements of the field are applied as per the standard fixed routine and distributed proportionately upon the crops as per defined rate and time. But most of the time this kind of techniques were not suitable for large area cultivation and lead to reduction on both quality and quantity of the crops yields.

First Consideration of real time variation of in-field soil health, crops conditions are very much essential for applying required amount of supplement types and most importantly to the specific site of the crops. Excessive application of crops nutrients not only wasted the resources, it also harmfully damages the production rate and quality of the crops. For the limiting resources, optimizing irrigation and crops management efficiency is very much in need at the current scenario. For regular changes in soil characteristics, difference in fertility: targeted pesticides distribution, localized manuring , site specific water application processes should be implemented on a real time situation. In this regards, recent advances in wireless communication technology and microcontrollers could help in adopting of real time wireless sensors network covering all the region of farm, specific target based water application, localized pesticide and manuring etc. Design of a reliable specific target based irrigation system is a challenging and complicated but will uplift the quality of agriculture products to some high extend [1].

II. MATERIALS AND METHODS

A. **Inventory of resources:**

In tea areas of northeast India, the source of water is usually rivers and perennial streams, shallow and deep tube wells. Usually surface water is preferred for economic reason; however, the discharge of the river or the stream must be sufficient to support the irrigation system even in the driest period of the year. If a dependable and easily accessible surface is not readily available, then the choice is to exploration of groundwater through installation of large diameter open wells or deep tube wells depending on the geomorphology of the area. The depth of the exploration tube wells can be approximately found out by conducting electrical resistivity test at the proposed site. In the low plain areas of the Brahmaputra valley, dependable yield of tube well for irrigation purpose can be expected within approximately 150m depths. However, towards the foothill areas depth of exploration increases as well as boundary substratum is encountered. Accordingly, exploration technique will also vary. Depending upon the area to be irrigated, number of irrigation sets will vary. It is always preferable to have tube well (in case of ground water source) centrally located, so that energy lost in overcoming friction in the mainline is reduced to minimum [2]. However, choice of power source is also another consideration. The function of the mainline is to supply water to the sprinklers from the source .Usually 125mm to 150mm diameter quick couple aluminum pipes or HDPE pipes of 6m standard lengths are used for the mainline [3].

B. *Selection of sprinklers and its spacing*

Rate of application of irrigation water must conform to the absorption rate of water by the soil, which is known as infiltration capacity. Excessive application rate shall result in wastage through runoff and slower application will be time and energy consuming, as well as area coverage will be also low [2].

The infiltration rate can be determined in situ by conducting double ring infiltrometer test. The approximate rate of application for most commonly available soil texture in the region is given in the table.1.

Table.1. Desirable rate of water application in mm/hour under various ground slopes

Ground Slope	**Soil texture**			
	Loamy sand	**Sandy loam**	**Silt loam**	**Clay loam**
0-5%	**20**	**13**	**10**	**8**
5-8%	**18**	**10**	**9**	**6**
8-12%	**15**	**8**	**7**	**5**
12-16%	**10**	**6**	**5**	**4**

Based on the infiltration rate, the size of nozzles, its operating pressure, and spacing between the sprinklers and between the laterals can be selected from the sprinklers' manufactures chart. An example is given in table 2. The efficiency of sprinklers depends upon the degree of uniformity of water application. The spacing and operating pressure regulate the uniformity of water application. Excessive force cause the droplet to break and lower pressure results in large droplet and smaller throw, and thus the uniformity of wetting the soil gets disturbed. The correct pressure of water application is ensured through monitoring of the nozzle pressure using a pressure gauge.

Table.2. Technical chart of sprinklers

Nozzle Diameter (mm)	Pressure kg/sq.cm	Diameter 'm'	Discharge lpm	Precipitation rate in mm/hr for spacing 30m x 30m
12x7	3.15 3.5	49.7 51.2	195 207	12.5 13.5
Premier ZM 22D	3.85 4.2 4.9 5.6	52.7 54.25 56.7 60.35	218 229 249 264	14.25 15 16.25 17.25
12x16 (HM106 PT)	4.5 5.5	48 56	248 273	16.53 12.64

C. *Wetting patterns*

The water distribution pattern from a single rotary sprinkler is not uniform. The area wetting pattern is circular typically, as shown in fig.1, and the dense wetting occurs close to the sprinkler. Therefore, to obtain uniformity, the patterns should overlap, as shown in fig.2, for which several sprinkler units must be installed very closely. Maximum spacing between sprinklers could be determined by providing an overlap of 60-65 percent of the wetted diameter in windless conditions [2]. If the wind velocity increases, then water distribution becomes distorted, and so in a high windy situation, the spacing of sprinklers should further be reduced.

In another case, to make the sprinkler work properly, it is necessary to provide the right operating pressure. Pressure directly relates to the distribution capacity of the sprinklers. Hence fluctuation in the pressure leads to an effect on wetted patterns of the sprinklers distribution.

Fig.1. Wetting pattern for a single sprinkler (TOP and SIDE view)

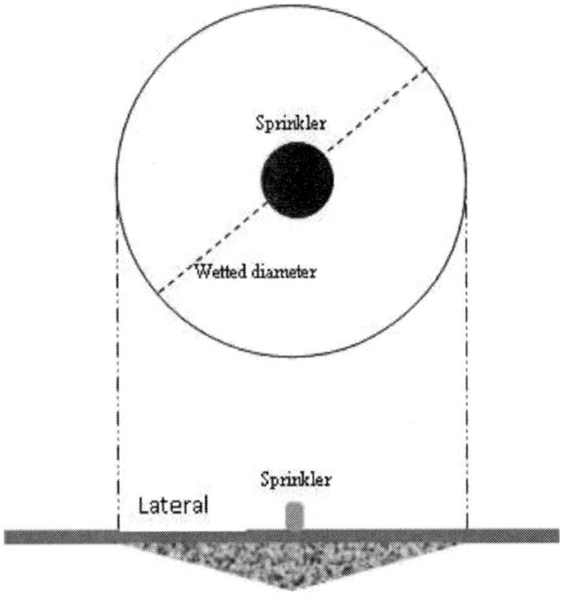

Fig.2. Uniform application of sprinkler

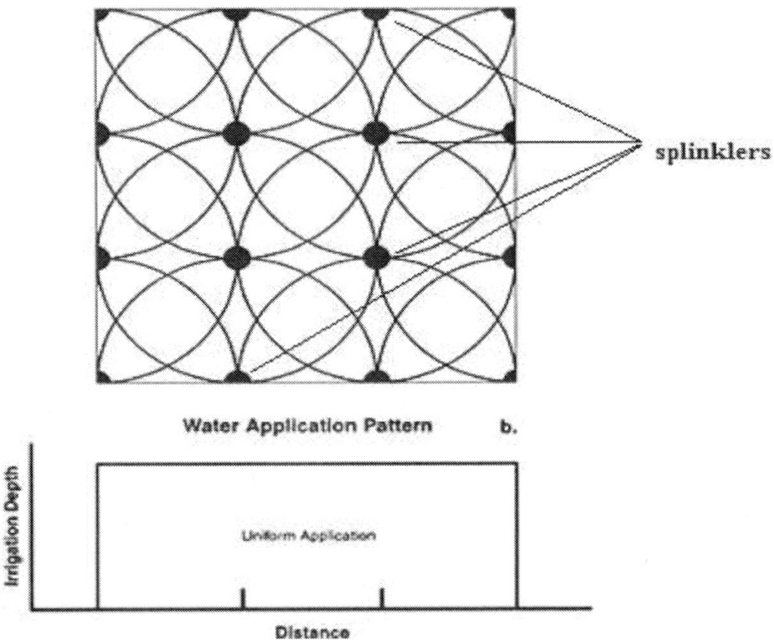

D. *Capacity of the Pump system*

For selecting the pump capacity, we need to consider the discharge of the well and the pump's energy to lift water against gravity to overcome various frictions in the system and create pressure at the nozzle head. The pump's energy requirement can be calculated from the system's layout and the type of nozzles selected.

$$\mathbf{Q = 27.8} \times \mathbf{A.d/F.H.E}$$

Q = discharge capacity of the pump in lit/sec
A = area to be irrigated, hectares
d = depth of application, mm

F = irrigation frequency, days
E = irrigation efficiency, percentage
H = operating hours per day, hr

The number of sets required to cover the desired area can work out accordingly. The pump should be chosen based on the energy needed to create to operate the system and its discharge. The system's total energy requirement is the energy required to overcome the lift of water from the source against gravity, head loss at entry, head loss at the exit, friction loss in the pipes and its joints, and the head required at sprinkler nozzles [2][4]. The energy (Horsepower) requirement of the pump can be worked out from the following relationship.

$$\mathbf{BHP = 1000\ Q\ H\ /75E}$$

Where, Q = discharge in liters per second (lps),
H = head in m,
E = Efficiency of pump

III. DESIGN AND IMPLEMENTATION OF THE EFFECTIVE IRRIGATION SYSTEM FOR TEA FARMING

A. ZIGBEE based Wireless node:

A wireless multi-sensing and actuator system (WSN) has been developed (Fig.3) for monitoring of different field parameters related to tea farming, can consist of ambient temperature, relative humidity, soil moisture, soil temperature, soil pH, solar radiation sensors, etc. [5] This embedded system mainly consists with a sensing unit, a user interface unit, a wireless or a wired communication unit and a data storage block. It has provisions for driving the actuation unit for controlling actions like sprinklers system in this case. The sensing module is sufficiently small in size, and it has the feature to interface multi-sensors along the single channel of ADC so that other ports can utilize in controlling others' processes [6]. The Implemented module has eight inputs of connection points for eight different sensors. Each sensor has a dedicated signal conditioning unit. No further signal conditioning is required externally. All necessary signals, conditioning, excitations, and linearization are embedded in the developed module.

In this system, each sensor unit is associated with a specific conditioning circuit. For the system's initial set-up, two push button switches are provided for the users' interface. At the start, the user must select the type of channel or sensor, and the channel's measurement ranges by setting the switch provided. The same system can be used for other classes and ranges of sensors with different configurations. For eight sensors input, there are eight different conditioning circuits. The control unit is consists of a microcontroller with ADC, a programmable gain amplifier (MCP6S28), and a power supply. The developed system is designed using an 8-bit microcontroller, atmega2560, which has 10 bits ADC with an acquisition time of 12.86ms, 256k bytes of flash, 4 Kbytes of data, and supports four programmable serial USART. Programmable gain amplifier (MCP6S28) gives the designer program control over an amplifier of the sensing section and can interface serially with the controller using the SPI bus.

For controlling the pressure of the Sprinkler system, the PWM control block is used. The stored set point from the microcontroller's internal memory (flash) can be select using a switch for each input parameter. The set value can be changed by using switches according to need. The program compares set values and sensing values than make the alarm sound if the sensing value exceeds the set point. Set points values can be adjusted later if needed through switches provided on the module by the user. Similarly, using corresponding switch set point value can be set for respective sensors inputs.

Fig.3: Wireless sensor node

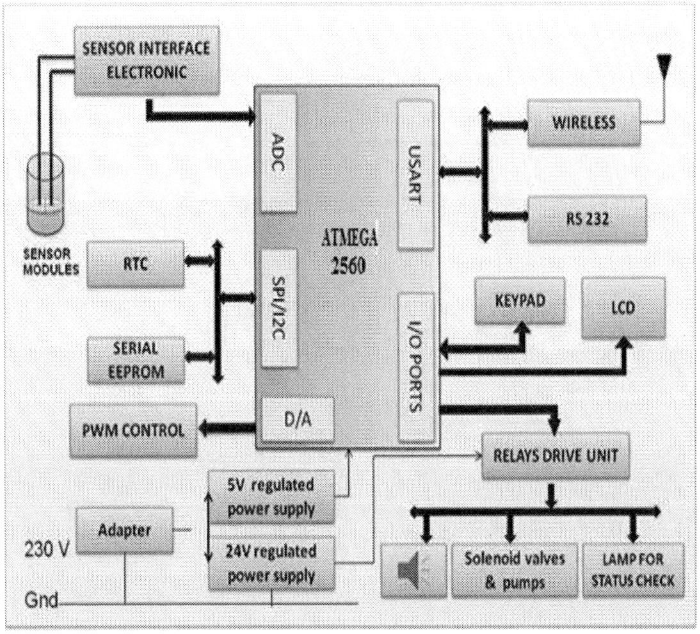

The alarm circuit is excited through an open collector source transistor array (ULN2003). Sensing data are stored in flash memory of the atmega2560 microcontroller and as well as on the serial EEPROM via I2C bus interface.

The wired module RS232 or unpaid wireless module Xbee transceiver, which has data transmitting capacity up to 1.6 km at 2.4G.Hz are used to transmit the measure parameter values and their set - point values and as well as to LCD unit simultaneously. The nodes operate under stored program control. The ADC embedded on the controller performs periodic scans of the sensors connected. The rate of scan is programmable and can be adaptive based on the changing rate of sensor reading. The acquisition time and conversion time should be enough fast to avoid any data loses during the real time interfacing of the system.

A. Field level wireless node deployment architecture

Simple field level deployment architecture of sensor nodes and decision support systems or units is shown in fig.4. Each wireless sensor nodes consist of multiple sensors, solenoid valves associated with sprinklers systems, a wireless module, and a microcontroller. As per the range of coverage of the wireless sensor module (here ZIGBEE RF module), the nodes are distributed along the farm area. For systematic and ease to access better to distribute nodes in a column-row matrix, preferably [6]. For comfort to localization, it is recommended to implements all the Sprinklers system laterally to the mainline pipe, and there may be 4-6 sprinklers in a node accordingly. The computational module on each sensor node is a programmable unit that performs computation, storage, and bi-directional communication with other sensor nodes. Each sensor node associated with the feedback control mechanism has been provided actuation unit and the sensing unit to allow control action transmitted from the decision support unit. Since the wireless communication range capacity of any RF wireless module is fixed, as an instance, the sensor node N23 act as coordinator node between node N21 and N24 in the implemented wireless sensor network. The coordinator node accumulates data from the two or more adjacent sensor nodes in a time-multiplexed manner, which helps in avoiding collisions of data transmissions [1].

Fig.4: A sample Field level WSN deployment architecture

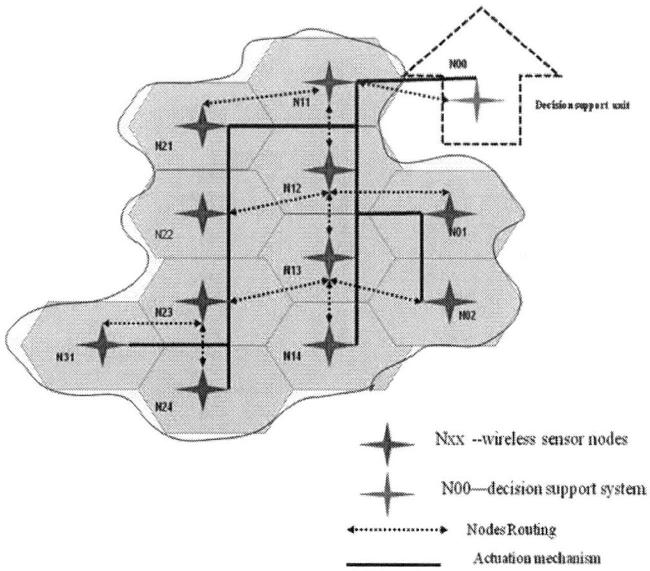

Along with its sensor data and data accompanied by the address of other neighbouring nodes aggregated, they are then transmitted to the next higher layer gateways nodes through a wireless link. The top-level sensor node transmits the sensor data received from the coordinators and from sensors to the remote computer terminal, i.e., decision support system node through the wireless interface. The computer terminal has the facility of data logging and decision support algorithms along with fast and powerful processing capabilities.

B. *Localized Sprinkler control using wireless node*

In this developed module, the XBEE-RF unit manufactured by DIGI is used to provide the required wireless communication link amongst sensor nodes and decision support unit. The modules are configured to meet IEEE 802.15.4 standards and provide self –organized, multi-hop, and reliable mesh networking with long power back up. The ZIGBEE protocol supports three types of nodes: coordinate, routers, and End devices.Figure.5 shows the ZIGBEE network topology.

Each ZIGBEE unit has two types of addresses, namely MAC address and Network address. The MAC address is 64 bits and is unique to each physical device, whereas the Network address is of 16 bits address, which was received randomly when a device joins a Zigbee network.

In this design, each node consists of a wireless unit (XBEE-RF), a microcontroller, sensor(s), actuator(s), Sprinkler(s) units, etc. All communication is performing using the source and destination 16-bit address. In the network routing tables to route data packets between the network can also use the 16 bits address, i.e., MAC address. However, this address is not static and can conflict between devices having the same 16-bit address. To avoid this phenomenon, in the implemented system, the unique 64-bits address is included in data transmission, ensuring to delivered data to the correct destination only. To control the targeted sprinklers units, the system must locate the unique 64-bits physical address of the sprinklers through the routing table. Accordingly, the controller has to excite the actuator of that associated node and make off the remaining undesired nodes sprinklers. A simple sprinkler control block is shown in fig.6.

Each wireless node was accompanied by one or more sprinkler systems and associated pressure gauge and solenoid valve to control the mechanism, as illustrated in the figure. All the control decisions were embedded on the Atmel 2560 microcontroller.

Fig.5. ZIGBEE network topology

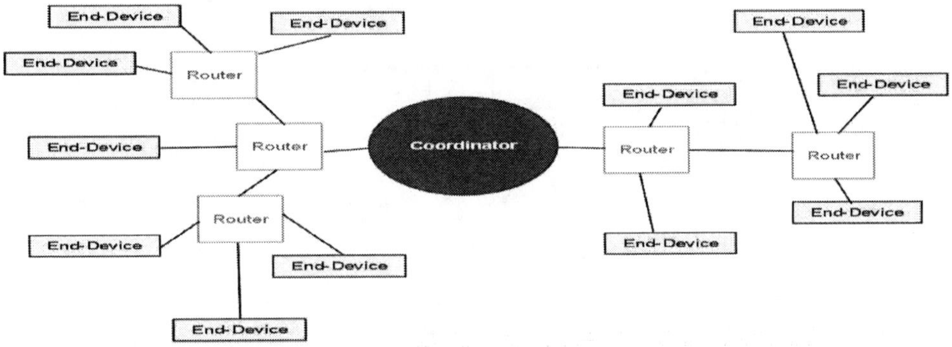

Fig.6. Sprinkler control block

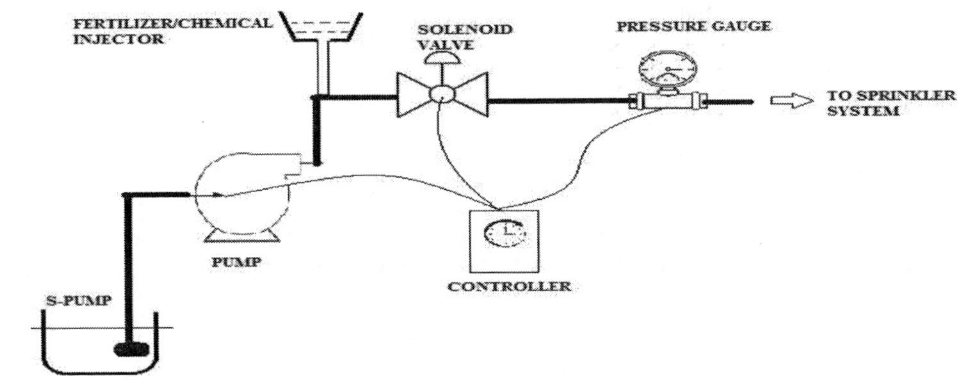

A. PWM based pressure control

The general control loop block diagram of the proportional valve implemented in the system's pressure pump is shown in fig.7. In this, the pressure sensor measures and transmits the current value of the process variable, PV, back to the controller. Controller error at current time t is computed as set point minus measured process variable. The controller uses this error in the control algorithm to calculate a new controller output signal, CO. Then the final controlled output is sent to the final control element that is pressure needle valve causing it to change. The variation in the final control element causes a change in the manipulated variable, which led to a change in the process flow rate.

Here, a PWM control unit has been implemented to set the set point as needed and avoids the sudden opening and closing of the proportional valve. In this, the PWM signal is generated by the microcontroller. The controller generates the corresponding 0- 2.4 kHz of PWM wave by the PWM module by configuring the respective register. The PWM signal from the microcontroller pin is again applied to a simple RC filter to convert it into a corresponding DC value.

Fig.7. Control Loop block.

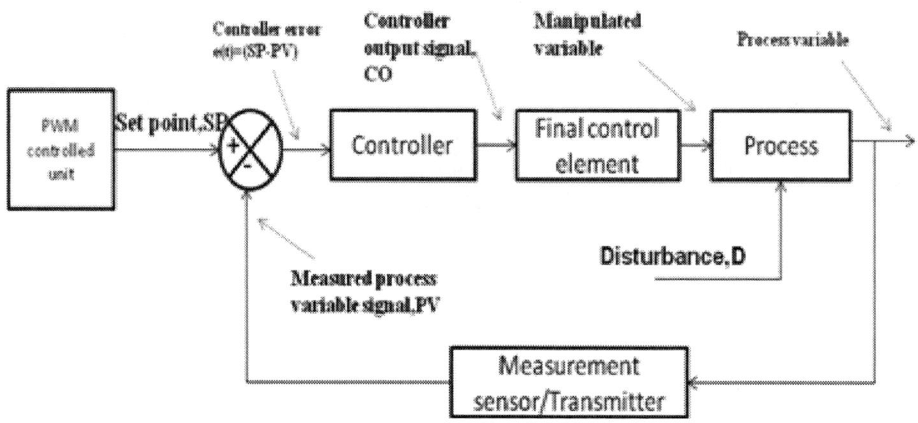

Since the proportional valve requires a control signal of 4-20 mA of current for closing and opening the pressure pump, the dc output from the RC filter is applied to the 4 -20mA converter IC (XTR110). Now, this can be used to control the proportional pressure valve.

E. Actuators drive unit

As mention in the previous section, the tea farm irrigation system's developed model includes two pumps and 4-6 solenoid valves for each node. So, to control these parameters, a relays drive unit has been designed along with a feedback control system. A simple relay control operation is shown in fig 9. In this microcontroller, control the relays through a darling ton transistor array IC, ULN2003.

Fig.8 Relay control

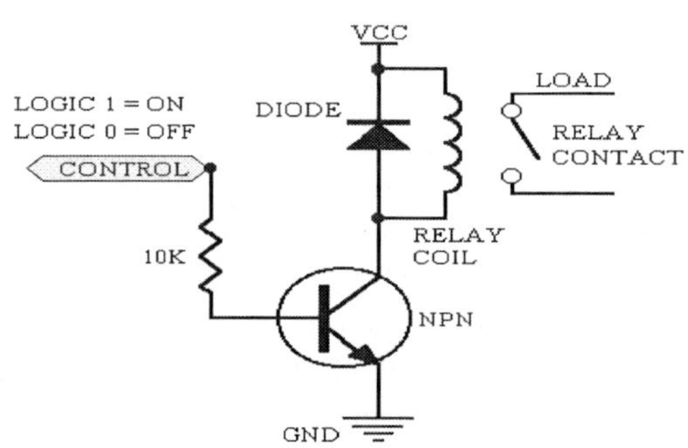

F. Feed back control

To protect from any unwanted operation in the node, it must be sure that each parameter should be excited at the respective time of each control action. Therefore, we need to check each parameter feedback status- just after sending the control action from the microcontroller that is before sending other parameters control actions.

So, a feedback control circuit has been added to the developed control system. Here, the Relays' feedback is taken from a connector and is fed to the input of a 4:1 MUX circuit (74LS153). The purpose of this multiplexer circuit is to minimize the I/O port assignment. The select lines of the MUX are connected to the two inputs pin of the microcontroller, respectively. According to the feedback control line's connection in the MUX input, we need to set the select line of the MUX so that we can check the status of that relay by checking the status of the microcontroller pin where the out of the MUX has connected.

IV. UNITS EXPERIMENTAL SET UPAND TESTING

A. Sensing node

For testing the developed system, an eight potentiometer circuit has a design in replica to eight sensors and is connected at the eight input of the programmable gain amplifier (PGA). Using the switches provided and following the instruction display on the LCD screen, set the programmable gain amplifier accordingly. The d.c 0-5Voutput of PGA is converted into digital form by the ADC. For the setting of the system, two push button switches are provided. Using the corresponding switch setpoint value can be set for respective sensors inputs. The alarm circuit is excited through an open collector source transistor array. For the transmission to remote personal computer or decision support unit, the microcontroller continuously checks if the character '#' has been received from the remote system unit; if received, then it transmits all measures parameters and corresponding sensors setpoint values to the hyper terminal and continues execution of the main program. The hyper terminal data transmission observed during the testing is shown in fig.9.

Fig.9. Hyper terminal data transmission of sensor node

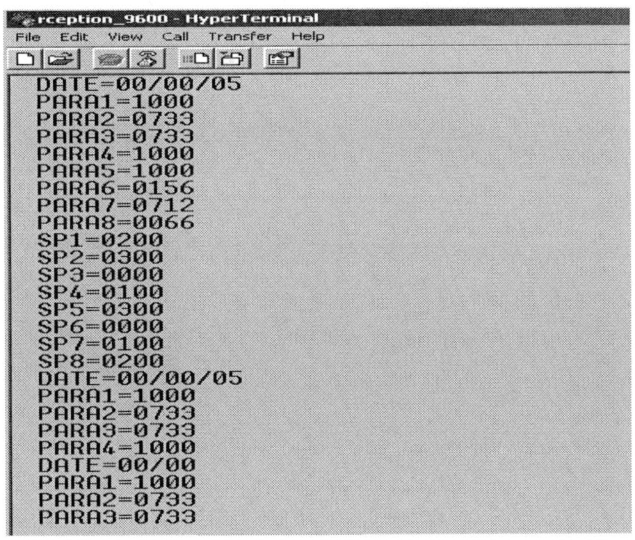

C. PWM control

To control the pressure of the system, a PWM based pressure control system has developed. It has been tested by applying different voltage levels to the ADC input and measured the corresponding control PWM generated by the microcontroller through algorithms. And now, check the output at the oscilloscope for each analog voltage level. The oscilloscope view of the generated PWM wave is shown in fig 10.

Fig.10: PWM wave with different applied input voltages

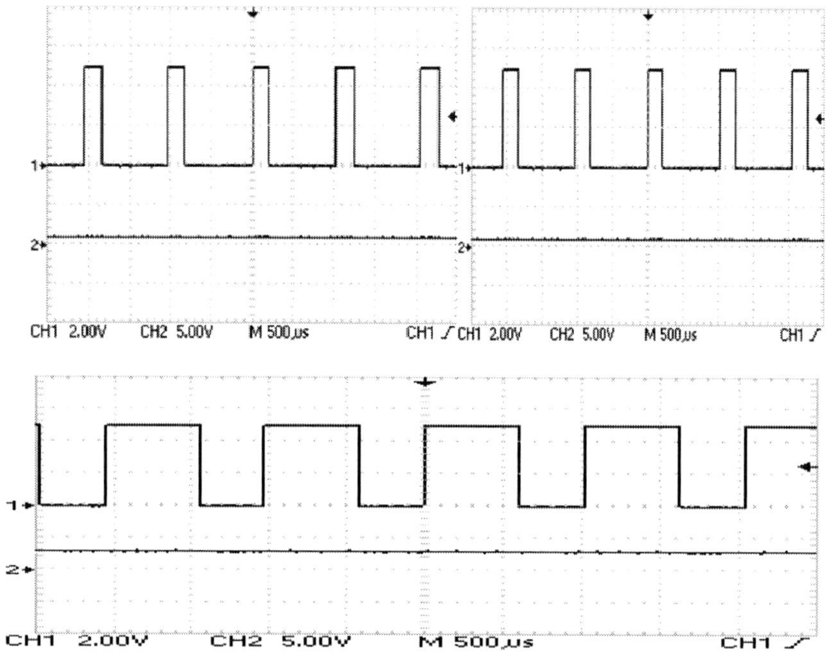

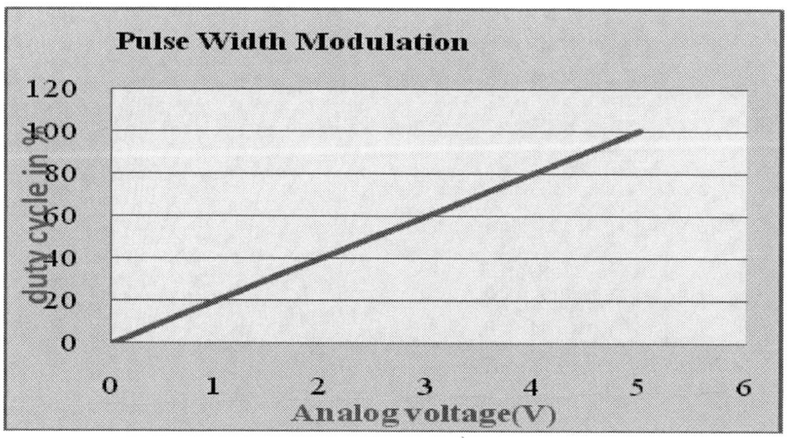

V. CONCLUSION

Large scale irrigation can be done by the method of flooding, furrowing, sprinkling, and drips. Because of the plant population density, the presence of a series of closely spaced open drainage systems, which debars practicing flooding irrigation, the sprinkler irrigation method is most popularly practiced in tea plantation throughout the world. It is suitable for all texture of soils found in the region as well for varied topographies. In this research, aiming to increase the quality and quantity of the crop product by utilizing minimum resources and reliable technologies, findings are performed. The developed wireless sensor network has performed Real-time field monitoring and control. Applications of farms' needs like fertilizers, pesticides, water, etc. are achieved through the sprinklers irrigation system. And most importantly, the intelligence of localized site –irrigation was embedded on the developed system using a unique 64 bits address identification methods. This system not only improves the quality and quantity of crops yield it also reduces the human effort and time consumes in farms testing, at a minimal cost.

References

[1] Xingye Zhu1, Prince Chikangaise,Weidong Shi, Wen-Hua Chen, Shouqi Yuan,"Review of intelligent sprinkler irrigation technologies for remote autonomous system" International Journal of Agricul-tural and Biological Engineering, (2018)11 (1), pp.23-30.

[2] Chen Dan, Ren Guang-wei, Wang Xiu-fang, Wang Xin-wei," Evaluation of Different Sprayers in the Tea Plants"2011 6th IEEE Conference on Industrial Electronics and Applications,pp.2403-2407

[3] Xiu-ji GUO*, Zheng FANG, Juan-juan CHEN, Jun-heng WANG,"Feasibility Study on Plastic Pipe Used in Simple Automatic Sprinkler System" 2014 7th International Conference on Intelligent Computation Technology and Automation.pp.260-263

[4] Lei Mai , Yongyi Cui , Fang Qu , Fan Wu,"Studies on water distribution characteristics of sprinkler in sprinkler systems"2015 Sixth International Conference on Intelligent Systems Design and Engineering Applications.pp.744-746.

[5] Channamallikarjuna, Mattihalli,Edemialem Gedefaye , FasilEndalamaw,AdugnaNecho,"Real time Automation of Agriculture Land, by Automatically Detecting Plant Leaf Diseases and auto Medicine" 2018 32nd International Conference on Advanced Information Networking and Applications Workshops.pp.325-330

[6] Sheng Jiang ; Weixing Wang , Guokun Xiao ; Zhen Li "Design of wireless sensor node for drought monitoring in tea plantation", 2011 International Conference on Electric Information and Control Engineering.pp-23-26.

[7] Winkler, D. A., Wang, R., Blanchette, F., Carreira-Perpinán, M., & Cerpa, A. E. (2016, April). MAGIC: Model-based actuation for ground irrigation control. In *2016 15th ACM/IEEE International Conference on Information Processing in Sensor Networks (IPSN)* (pp. 1-12). IEEE.

[8] Sung, W. T., Chen, J. H., Hsiao, C. L., & Lin, J. S. (2014, June). Multi-sensors data fusion based on arduino board and xbee module technology. In *2014 International Symposium on Computer, Consumer and Control* (pp. 422-425). IEEE.

[9] Parmenter, J., Jensen, A. N., & Chiu, S. (2014, June). Smart irrigation controller. In *IEEE International Conference on Electro/Information Technology* (pp. 394-398). IEEE.

[10] Sales, N., Remédios, O., & Arsenio, A. (2015, December). Wireless sensor and actuator system for smart irrigation on the cloud. In *2015 IEEE 2nd World Forum on Internet of Things (WF-IoT)* (pp. 693-698). IEEE.

[11] Ningombam, H., & Roy, O. P. (2019, August). A Systematic Approach on Design of Automation System for Tea Farming Based on Embedded Wireless Sensor Network. In *International Conference on Inventive Computation Technologies* (pp. 286-296). Springer, Cham.

[12] Jha, K., Doshi, A., Patel, P., & Shah, M. (2019). A comprehensive review on automation in agriculture using artificial intelligence. *Artificial Intelligence in Agriculture*, *2*, 1-12.

[13] Aoki, E., Kudo, K., Fukuda, A., Nakanishi, T., Tagashira, S., Okayasu, T., ... & Imura, Y. (2012, February). Knowledge management platform based on the environmental monitoring system with energy harvesting sensor motes for tea farming. In *Proc. of 2012 International Conference on Affective Computing and Intelligent Interaction (ICACII2012), Lecture Notes in Information Technology* (Vol. 10, pp. 217-223).

[14] Mansour, H. A., Pibars, S. K., Abd El-Hady, M., & Ebtisam, I. E. (2014). Effect of water management by drip irrigation automation controller system on faba bean production under water deficit. *International Journal of GEOMATE*, *7*(2), 1047-1053

[15] THEREZA, N., SAPUTRA, I. P. A., & HAMDADI, A. (2020, May). The Design of Monitoring System of Smart Farming Based on IoT Technology to Support Operational Management of Tea Plantation. In *Sriwijaya International Conference on Information Technology and Its Applications (SICONIAN 2019)* (pp. 52-57). Atlantis Press.

[16] Romanov, V., Gribova, V., Galelyuka, I., Voronenko, O., Abdel-Maboud, N. F., ELBagoury, B. M., ... & Lytvynov, V. (2015). Multilevel sensor networks for precision farming and environmental monitoring. *International Journal" Information Technologies & Knowledge" ITHEA.–Sofia, Bulgaria*, *9*(1), 3-10.

[17] Ahuja, P. S., Gulati, A., Singh, R. D., Sud, R. K., & Boruah, R. C. (2013). *Science of Tea Technology*. Scientific Publishers.

[18] Munene, N. K., & Kasamani, B. S. (2018). A Mobile-based TeaF arming Information System. *International Journal of Computer Applications*, *975*, 8887.

[19] Chan, Y., Hu, J., Chou, C., Liao, C., & Chen, C. (2017). Development of ICT for Leaching Monitoring in Taiwan AgriculturalLTER tations. *Environments*, *4*(3), 47.

[20] Keshtgary, M., & Deljoo, A. (2012). An efficient wireless sensor network for precision agriculture. *Canadian Journal on Multimedia and Wireless Networks*, *3*(1), 1-5.

[21] Mafuta, M., Zennaro, M., Bagula, A., Ault, G., Gombachika, H., & Chadza, T. (2013). Successful deployment of a wireless sensor network for precision agriculture in Malawi. *International Journal of Distributed Sensor Networks*, *9*(5), 150703.

[22] Srbinovska, M., Gavrovski, C., Dimcev, V., Krkoleva, A., & Borozan, V. (2015). Environmental parameters monitoring in precision agriculture using wireless sensor networks. *Journal of cleaner production*, *88*, 297-307.

[23] Sahitya, G., Balaji, N., Naidu, C. D., & Abinaya, S. (2017, January). Designing a wireless sensor network for precision agriculture using ZigBee. In *2017 IEEE 7th International Advance Computing Conference (IACC)* (pp. 287-291). IEEE.

[24] Kaewmard, N., & Saiyod, S. (2014, October). Sensor data collection and irrigation control on vegetable crop using smart phone and wireless sensor networks for smart farm. In *2014 IEEE Conference on Wireless Sensors (ICWiSE)* (pp. 106-112). IEEE.

[25] ZHANG, X. L., DING, F., ZHANG, S. Q., & YANG, J. (2007). Study on the Wireless Data Acquisition System under Greenhouse Environment [J]. *China Rural Water and Hydropower*, *2*.

Mentoring and Career Success among middle-level Managers: Analyzing the Mediating role of Emotional Intelligence

Mudang Tagiya

Department of Humanities and Social Sciences (HSS), NERIST, Nirjuli, Arunachal Pradesh, India – 791109

e.mail: rs_mudang_hss15@nerist.ac.in , tagiyam@gmail.com

Shyamalee Sinha

Department of Humanities and Social Sciences (HSS), NERIST, Nirjuli, Arunachal Pradesh, India – 791109

e.mail: ss@nerist.ac.in

Passang Norbu

Department of Commerce, Rajiv Gandhi University, Doimukh, Arunachal Pradesh, India – 791112

e.mail: passangnorbu7@gmail.com

Abstract. Mentoring is proven to be a significant factor in building one's career. Every successful story build under various underlying elements where mentoring is count as one of those. Similarly, there are many predictors that are responsible for one's success personal as well as professional. It is very possible that these predictors may have interrelations and tendency to influence directly or indirectly. Therefore, the study attempt to study the influence of emotional intelligence as mediating variable between mentoring and subjective career success among the mid-level managers working in the financial institutions (nationalized banks) in the state of Arunachal Pradesh, India. A structured questionnaire was used as an instrument for data collection which was carried out through face to face surveys and online surveys via. Google forms. The respondents for the study were mid-level managers (n=252) working in a nationalized bank in the state of Arunachal Pradesh. The software package viz. SMART PLS 3.0 (trial version) and SPSS version 21 were used for data analysis and mediation analysis. The output shows that Emotional intelligence has a mediating effect on mentoring and career success. The very nature of mediation effect of emotional intelligence was found to be full mediation. As much as predictive relevance of the obtained model, the results shows that the derived mediation model has a medium predictive relevance. The outcome of this study will add value to the multifaceted nature of the relationship of mentoring with career success.

1. Introduction

Most scholars who have given particular attention to career selection issues of all staff and organizations in the area of human resources administration have concentrated intensely and studied thoroughly. [1] have especially discussed the dynamic nature of productive employment and having a background of career success and its interactions will help human resources administrators appreciate career management as a method. This paper further extends their work and analyses the relations of career success and mentoring of middle-level managers working in nationalized banks in the study area.

1.1. Mentoring

Mentoring is defined as a "reciprocal relationship with an intentional agenda, designed to communicate explicit content along with life wisdom from one individual to another [2]. It is clear from the above definitions that mentoring does not happen by accident. Mentoring is acknowledged as a vital developmental source for people in organizations [3] and has been described as the most powerful and influential one-on-one progressive affiliation [4]. Scholars from different time argued that mentoring is of utmost importance for the progress of human capitals [5]; [6]. [7] in their work found that mentoring has impacted significantly in personal learning. [8] mentions the work by [9] and [10] which provokes for following research in the area. Both the writers projected it was advantageous to a protégé's outcome due to access to a mentor. [10] added it to be a significant growing development in majority. [9] observed that apart from desirable occupations for wards, mentoring also enables them to the authority structures in the organizations. Furthermore, he associated with rejuvenation and satisfaction to the mentor as certain benefits of assisting others. Scholars like [11] also addressed the adult development for both mentee and mentor through mentoring facilitates whereas [12] addresses the psychological development of both mentoring partners. To add, scholars like [13] and [10] states that the connection acts as a prospect for mentors. Apart from these, since the forthcoming leaders are equipped through mentoring, the organization also benefits out of it [14]. Further, commitment and satisfaction are also found for those employees who are mentored [15]. Also, mentoring adds significantly in higher pay [16], salary progress and promotion rate [23] and career fulfilment [17].

1.2. Career Success

Career success has been defined as "the positive psychological or work-related outcomes or achievements that one has accumulated as a result of one's work or work experience" [18]; [1]. Success includes both the real and perceived achievements individuals have accumulated as a result of their work experience [1]. Due to the positive outcomes like salary level, promotion, job satisfaction etc. career success has become a significant in one's life [19]. It is also critical for an organization since productive employees are able to add value which influences organization efficiency. [20] It has contributed to a large degree of curiosity in its forecast.

1.3. Mentoring and Career Success

Mentoring has been related to career progress and development by many scholars [21] [22] [24] [25] [26] [27]. The findings of [28] suggests that there is a key role played by mentoring in the career development in senior management positions in the NHS Scotland. Mentoring has also been associated to career success, [29] and is confirmed by scholars to be beneficial for career success [30] [31]

H1: *Emotional intelligence mediating in the relationship between mentoring and subjective career success of mid-level managers.*

1.4. Mentoring and Emotional Intelligence

Good mentors are socially self-aware and self-regulating, recognize how their attitude impacts on their colleague & subordinates, and have empathy, understanding, and sincere engagement in helping others. Such aspects of emotional intelligence are particularly relevant in order to successfully mentor among genders, generations and relationships. Emotional intelligence is critical for a mentor. The association between emotional intelligence and the degree of faith a mentor has in them was favourable among the mentors [32]. There are also several skills that a trainer or instructor may usefully build. Strong empathy seems to enhance brighter career and mental support [32]. It is necessary to measure EI, explored, and both mentor and mentee to develop self-awareness of their respective emotional intelligence to ensure optimum result from mentor-mentee relationship [33]. This synergy between both will enrich learning, strengthen interpersonal relationship, retention and improve HRD. [34] suggested the elements of empathy, physical awareness, and relationship building to be incorporated in the training along with emotional connection for better engagement, change management, resilience, leadership and motivation among trainee. Emotions architect social relationship and, as such, a mechanism like mentoring would tend to be viewed via. prism of emotions, with interpersonal and social contacts at its heart [35].

H1a: There is a significant relationship between Mentoring and Emotional intelligence among mid-level managers.

1.5. Emotional Intelligence and Career Success

Emotional Intelligence is comprised of several key components as outlined by [36]. According to them, it is "the ability to understand, recognize, and express emotions in oneself; the ability to understand and empathize with the emotions of others; the capacity to manage and control emotions in oneself and others; the ability to manage change, adapt to changing environments and conditions; and the ability to use the knowledge of emotional states to solve problems and make decisions." An individual's level of emotionality has an effect on their workplace experiences and career decisions [37] [38]. It has also been suggested that when emotions are well understood and make good use of it, it positively influences the successful career [39] [40]. Similar findings have been resulted by [41] which indicated that emotional intelligence and career adaptability have positive relationship. Studies have shown that people with high levels of emotional intelligence have greater career success, foster stronger personal relations, have effective leadership skills and are healthier than those with low emotional intelligence [42]. According to [43], emotional intelligence is an important aspect in shaping career success while [44] believes it stimulates how people engage in workplace situations and career decision making. Another interesting finding has been presented by [45] where they stated better prepared to take care ambiguity and on-the-job stressors are the ones with high EI. The study conducted on mid-level service managers in the least density populated state in India, analyzing the relationship between emotional intelligence and subjective career success has also found significant positive relation [46]. The result obtained were inconsonance with earlier similar studies.
Thus, through existing literature, it may be concluded that having a strong emotional intelligence is crucial for career advancement & success and to be fruitful in the workplace.

H1b: There is a significant relationship between Emotional intelligence and career success among mid-level managers. `

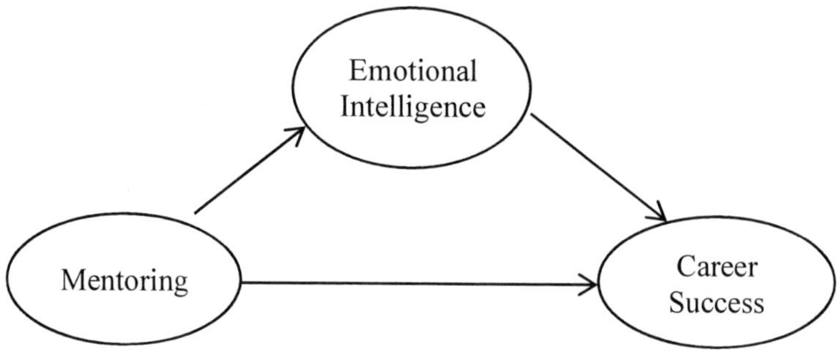

Figure 1. Conceptual model of the study

Hence, the present study aims to explore the relationships between the three variables.

2. Research Methodology

2.1. Data Collection

In this study, a questionnaire was used for data collection. The data were collected for a period of 6 months i.e. from July 2019 – January 2020. The sample respondents were mid-level Managers working in financial institutions, i.e., nationalized banks and NBFCs operating in the state of Arunachal Pradesh, India. Based on the quota sampling method, 300 (Three hundred) samples were surveyed out of which 252 (two hundred fifty-two) were selected after eliminating 48 respondents who failed to respond or responded inappropriately. The adopted scales incorporated with demographic questions in the form of an online questionnaire and hard copies were used to collect the data. The data was collected through face to face surveys and email surveys via. Google-form.

2.2. Sample characteristics

In the demographic profile of the sample respondents, the majority of the sample respondents (N=252) range form 26-35 years of age. In terms of gender, 38 respondents were female and 214 were male. Over half (54%) had completed their graduation, 38.9% held a post-graduation, 4.8 % held diploma, 1.6% held M.Phil. About 60% (n=152) of the respondents were having working experience in the range from 6 - 10 years, in their field, followed by the range 11-15 years (n=70). In terms of income level, 29.9% of the respondent have the income level of range from 75,001 - 1, 00,000 (INR) followed by 27% from the range of 1, 00,001 - 1, 25,001 (INR).

2.3. Research Instrument

The research instrument (questionnaire) contains 33 questions which were classified into two parts. The first part covers the demographic information which contains 8-question e.g., name, age, income, experience, educational qualification, etc. The second part of the instrument covers the variables to be measure for the study which consists of 26 questions of interval scale that are designed based on 5-point Likert Scales.

2.3.1. Measurement, reliability, and validity of scales

To ensure the validity of the research instrument, the items in the instrument for the study were drawn from various authentic sources under the guidance of experts. For the reliability of the whole instrument the scale reliability was conducted and the results reveal that the Cronbach's alpha of the instrument was found to be above-accepted value ($\alpha > 0.7$).

2.3.2. Subjective career success

Subjective career success' was measured via. 'career satisfaction'. Items of the scale were rated on a 5-point Likert Scale (1= Strongly Disagree; 5= Strongly Agree). [47] five-item measure was used to measure '*career satisfaction'*, sample items: "I am satisfied with the progress I have made toward meeting my overall career goals", "I am satisfied with the progress I have made toward meeting my overall career goals", "I am satisfied with the progress I have made toward meeting my goals for income", "I am satisfied with the progress I have made toward meeting my goals for advancement". The measuring scale for subjective career success was proved to be reliable with Cronbach's $\alpha = .88$.

2.3.3. Emotional Intelligence

'Emotional intelligence' was measured by the '10-item "Breif Emotional Intelligence Scale (BEIS – 10)" retrieved from [48]. Items of the scale were rated on 5-point Likert Scale (1= "not at all characteristic"; 5 = "very characteristic"). The sample items were; "*I know why my emotions change*", "*I easily recognize my emotions as I experience them*", "I seek out activities that make me happy". The measuring scale for emotional intelligence was proved to be reliable with Cronbach's $\alpha = 0.92$.

2.3.4. Mentoring

To measure mentoring, we used the 11-item scale developed by [49] mentoring scale. The scale measures two different dimensions of mentoring i.e. career development and psychological support. The first 6-items of the scale measures the 'career development' aspect and the next 5-items measure 'psychological support' aspect of mentoring. Items of the scale were rated on a '5-point Likert Scale' (1= 'Strongly Disagree'; 5= 'Strongly Agree'). Mentoring was operationalized as an average of 11-items. The sample items of the scale are: *"My mentor takes a personal interest in my career", "My mentor gives me special coaching on the job", "I shared personal problems with my mentor", "I exchange confidence with my mentor"*. The measuring scale for mentoring was proved to be reliable with Cronbach's $\alpha = 0.91$.

2.3.5. Demographic variables

At the end of the questionnaire, some demographic information were asked which included 'gender' (1 = '*male'*, 2 = '*female'*), 'age', 'educational level' (1 = '*upto higher secondary level'*, 6 = Ph.D.) 'current working experiences' (1 = '*upto 5 years'*, 5 = '*more than 20 years'*), 'monthly income' of the family (1= '*upto 75,000'* INR, 7= '*more than 2 lakh'* INR).

3. Results and Discussion

3.1. Descriptive statistics

The descriptive statistics (mean and standard deviation) and correlations among the study variables are presented below in Table no. 1. Results in Table-1 showed that mentoring was positively correlated with Emotional Intelligence and Career Success (0.610** & 0.404**, $p < 0.01$) Furthermore, it was found that "Age" was negatively associated with "Gender" (-0.216**, $p<0.01$) and positive relation with Emotional Intelligence (0.127**, $p<0.05$). Finally, results depicted in Table 1 also reveals that age was positively linked to career success also (0.191*, $p<0.01$), however, educational qualification had no significant association with any of the study variables.

Table 1. Mean, Standard deviation and correlation

	M	SD	1	2	3	4	5	6
1. Gender	1.15	.36	1					
2. Age	2.48	.73	-.216**	1				
3. Education Qualification	3.40	.65	.033	-.093	1			
4. Mentoring	3.21	.708	-.021	.065	-.037	1		
5. Emotional Intelligence	3.84	.616	-.001	.127*	.003	.610**	1	
6. Career Success (CS)	3.63	.723	-.013	.191*	-.095	.404**	.538**	1

N=252

**0.01 level (1-tailed).

* 0.05 level (1-tailed).

For the mediation study, we have checked the significant relationship between the independent and dependent variables of the study. The path model was assessed via bootstrapping without the interaction of a mediator (Figure 2). The results show that the relationship between mentoring and career success is found to be significant statistically (t-value = 8.415, p<0.05, subsample: 50000).

Table 2. Path Analysis (Direct effect)

Path	*B*	*Std. error*	*t-values*	*p-value*	*Decision*
Mentoring → Career Success	0.179	0.05	8.415	.000*	Accepted

*p < 0.05 (1-tailed)

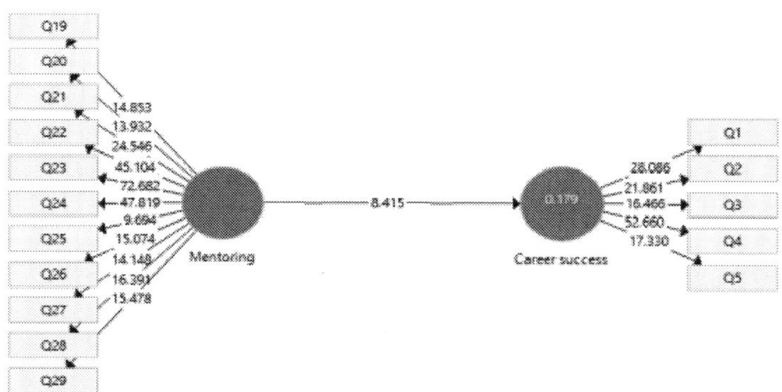

Figure 2: Direct effect between Dependent variable and Independent variable

3.2. Measurement Model

The measurement model consisted of three latent variables: Subjective Career Success (CS), Emotional Intelligence (EI), and Mentoring. Career success (5- items), Emotional Intelligence (10-items), and mentoring (11- items) were modeled by their respective items. The result obtained from the smart PLS shows that the convergent validity and discriminant validity of the data set for the variable was adequate (Alpha >0.7, Roh-A> 0.8, CR>0.7 and AVE>0.5). Table no. 2 highlights the details measurement model.

Table 3. Reliability and AVE

Construct	*Loading*	*No. of Items*	*Alpha*	*Roh A*	*CR*	*AVE*
Career Success	0.653 - 0.831	05	0.864	0.877	0.867	0.570
Emotional Intelligence	0.601 - 0.824	10	0.917	0.927	0.921	0.542
Mentoring	0.606 - 0.874	11	0.918	0.928	0.918	0.512

Source: Smart PLS 0.3 output

3.3. Structure Model

Henceforth, the inclusion of the variable viz. emotional intelligence as a mediator was meaningful enough to proceed further. For the mediation, we require the addition of indirect paths to verify that Emotional intelligence mediates the relationship between mentoring and career success. To test the model and hypothesis, the path model was assessed via bootstrapping, without the interaction of a mediator (Figure 3).

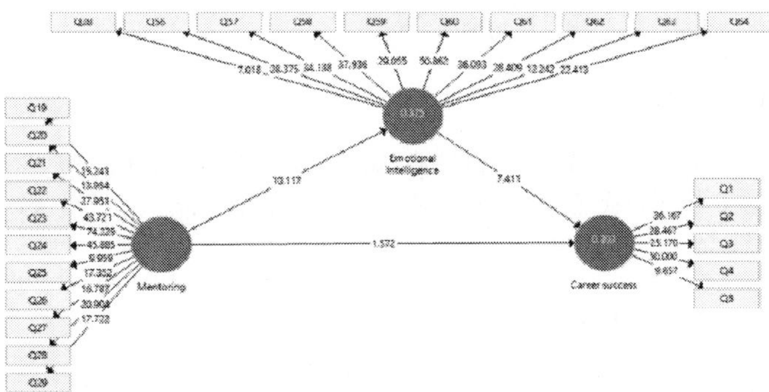

Figure 3: In-direct Effect between the variables (Bootstrapping)

The t-value of the indirect path "Mentoring→ emotional intelligence" was 10.116 with a p-value of 0.00* at 5% significant and path "Emotional intelligence → Career success" was 7.411 with a p-value of 0.00* at 5% significant (Table, 4). However, the direct path between mentoring and career success was found to be insignificant with t = 1.52 and p = 0.058. Since direct relation found to be insignificant, it is interpreted that the obtained model is of full mediation.

Table 4. Path analysis for median effect

Path	***B***	**S.D**	**t -statistics**	**p values**	**Inference**
EI → CS	0.556	0.065	7.411	0.000*	Significant
Mentoring → CS	0.073	0.064	1.572	0.058	Not Significant
Mentoring → EI	0.660	0.061	10.116	0.000*	Significant

*p<0.05, EI : Emotional Intelligence, CS : Career Success

Hence, the presence of emotional intelligence in between mentoring and career success has a full or complete mediation between the variables in the obtained model. Consequently, we fail to reject the null hypothesis. Hence, the proposed hypothesis is supported and accepted.

Table 5. Testing of model and PLS output

Set/model	Independent variable	Mediating variable	Dependent variable	PLS output	Hypothesis results
Structure model	Mentoring	Emotional Intelligence	Career success	R2= 0.367, Q2 = 0.19, f2 = 0.771, 0.21	Supported

Source: Smart PLS 3.0 output

Further, the blindfolding was performed in the model with omission distance of five (table no. 5), which reveals that the predictive relevance (Q2) of the model has a medium predictive relevance from the study (i.e. Q2 = 0.19) and effect size (f2) between Mentoring– Emotional intelligence and Emotional intelligence – career success was 0.771 and 0.27 which is significant and hence hypothesis results are supported.

4. Scope for further studies

The present study attempts to understand the mediating role of Emotional Intelligence between Mentoring and Career Success. Since there was one mediating variable so the study was simple mediation analysis. There are various predictors associated with career success and one can undertake higher level mediation study like parallel mediation, serial mediation, moderating mediation analysis etc. to understand the complex picture of career success.

5. Conclusion

To understand the degree of mediation effect in this empirical analysis, a model was hypothetically examined and attempt has been made to explain it through an illustrative model. Results obtained from the initial analysis i.e. table 3 showed the significance of the direct path between the dependent and independent variables. Further, the significant indirect paths were found in *table 4*. The presence of mediating variables in the between independent and dependent variables affect the direct effect between them and led to the relation insignificant resulting in the full mediation effect of the mediating variable i.e. emotional intelligence. Finally, predictive relevance and effect size of the model was medium and found (R2= 0.367, CV-Redundancy= 0.19, f2= 0.771, 0.21) to be significant i.e. effect of mentoring on career success is explained via Emotional intelligence. Given this, it can be assumed that emotional intelligence has a full mediation and plays a significant role in between mentoring and career success. Hence, it can be concluded that career success can be strengthened and enhanced by the addition of emotional intelligence among mid-level managers.

References

[1] Judge, T. A., Cable, D. M., Boudreau, J. W., & Bretz Jr, R. D. (1995). An empirical investigation of the predictors of executive career success. *Personnel psychology*, *48*(3), 485-519.

[2] Addington T & Graves S (2017). *Training Journal.* Retrieved from http://www.trainingjournal.com/blog/16-mentoring-definitions

[3] Noe, R. A., Greenberger, D. B. and Wang, S. (2002). Mentoring: What we know and where we might go. In G. R. Ferris (Ed.), *Personnel and human resources management,* 21: 129-173

[4] Wanberg, C. R., Welsh, E. T., and Hezlett, S. A. (2003). Mentoring research: A review and dynamic process model. In J. J. Martocchio & G. R. Ferris (Eds.), In *Research in Personnel and Human Resources Management* (pp. 39-124). (Research in Personnel and Human Resources Management; Vol. 22).

[5] Kram, K. E. (1988). Mentoring at work: Developmental relationships in organizational life. *University Press of America.*

[6] Bozeman, B., and Feeney, M. K. (2007). Toward a useful theory of mentoring: A conceptual analysis and critique. *Administrative and Society*, 39(6), pp. 719-739.

[7] Scandura, T. A. and Lankau, M. J. (1996). Relationships of Gender, Family Responsibility and Flexible Work Hours to Organizational Commitment and Job Satisfaction. *Journal of Organizational Behaviour*, vol. 18, pp. 377-391.

[8] Ebrich, L. C. and Hansford, B. (1999). Mentoring: Pros and Cons for HRM. *Asia Pacific Journal of Human Resources*, 37(3), 1999.

[9] Kanter, R. (1977). Men and women of the corporation. New York: *Basic Books.*

[10] Levinson, D. J., Darrow, C. M., Klein, E. G., Levinson, M. H., & Mckee, B. (1978) The seasons of a man's life. New York: *Knopf.*

[11] Merriam, S. (1983). Mentors and protégés: A critical review of the literature. *Adult Education Quarterly*, 33, pp. 161-173.

[12] Bartell, C. A. (2005). Cultivating high-quality teaching through induction and mentoring. *Thousand Oaks*, CA: Corwin Press.

[13] Dalton, G. W., Thompson, P. A., and Price, R. L. (1977). The four stages of professional careers- a new look at performance by professionals. *Organizational Dynamics*, 6, pp. 19-42.

[14] Zaleznik, A. (1977). Managers and leaders: are they different? *Harvard Business Review*, 55, 67-78.

[15] Hunt, D. M., and Michael, C. (1983). Mentorship: a career training and development tool. *Academy of Management Review*, 8, pp. 475-485.

[16] Roche, G. R. (1979). Much ado about mentors. *Harvard Business Review*, 57, pp. 14-28

[17] Burke, R. J. (1984). Mentors in organizations. Group and Organization Studies, 9, pp. 253-272.

[18] Arthur, M. B., Khapova, S. N., & Wilderom, C. P. (2005). Career success in a boundaryless career world. *Journal of Organizational Behavior: The International Journal of Industrial, Occupational and Organizational Psychology and Behavior*, *26*(2), 177-202.

[19] Judge, T. A., Higgins, C. A., Thoresen, C. J., & Barrick, M. R. (1999). The big five personality traits, general mental ability, and career success across the life span. *Personnel psychology*, *52*(3), 621-652.

[20] Delaney, J. T., & Huselid, M. A. (1996). The impact of human resource management practices on perceptions of organizational performance. *Academy of Management journal*, *39*(4), 949-969.

[21] Lunding, F. S., Clements, C. E., and Perkins, D. S. (1978). Everyone who makes it has a mentor. *Harvard Business Review*, 56, pp. 89-101.

[22] Kram, K. E., and Isabella, L. A. (1985). Mentoring alternatives: The role of peer relationships in career development. *Academy of Management Journal,* 28(1), pp. 110-132.

[23] Scandura, T. A. and Schriesheim, C. A. (1994). Leader-Member Exchange and Supervisor Career Mentoring as Complementary Constructs in Leadership Research. *Academy of Management Journal*, 37, pp. 1588-1602.

[24] Scandura, T. A. (1992). Mentorship and career mobility: an empirical investigation. *Journal of Organizational Behavior*, 13, pp. 169-174.

[25] Gubbins, M. C., and Garavan, T. N. (2005). Studying HRD practitioners: A social capital model. *Human Resource Development Review*, 4(2), pp. 189-218.

[26] Briggs, E., Jaramillo, F., and Weeks, W. A. (2011). Perceived barriers to career advancement and organizational commitment in sales. *Journal of Business Research,* 65, pp. 937-943.

[27] Durbin, S., and Tomilson, J. (2014). Female part-time managers: Careers, mentors and role models. *Gender, Work and Organization,* 21, pp. 308-320.

[28] Seraj, S. B., Tsouroufli, M. and Branine, M. (2015). Gender, Mentoring and Social Capital in the National Health Service (NHS) in Scotland, UK. *International Perspectives on Equality, Diversity and Inclusion*, Volume 2, 25 Sept. 2015, pp. 43-68.

[29] Darwin, A. (2000). Critical reflections on mentoring in work settings. *Adult Education Quarterly*, 50, pp. 197-211.

[30] Allen, T. D., Eby, L. T., Poteet, M. L., Lentz, E., and Lima, L. (2004). Career benefits associated with mentoring for proteges: A meta-analysis. *Journal of Applied Psychology*, 89(1), pp. 127-136.

[31] O'Brien, K. E., Biga, A., Kessler, S. R., and Allen, T. D. (2010). A meta-analytic investigation of gender differences in mentoring. *Journal of Management*, 36(2), pp. 537-554.

[32] Chun, J. U., Litzky, B. E., Sosik, J. J., Bechtold, D. C., & Godshalk, V. M. (2010). Emotional intelligence and trust in formal mentoring programs. *Group & Organization Management*, *35*(4), 421-455.

[33] Opengart, R., & Bierema, L. (2015). Emotionally intelligent mentoring: Reconceptualizing effective mentoring relationships. *Human Resource Development Review*, *14*(3), 234-258.

[34] Cremona, K. (2010). Coaching and emotions: An exploration of how coaches engage and think about emotion. *Coaching: An International Journal of Theory, Research and Practice*, *3*(1), 46-59.

[35] Fineman, S. (Ed.). (2000). *Emotion in organizations*. Sage.

[36] Bar-On, R., Handley, R., and Fund, S. (2013). The impact of emotional intelligence on performance. In V. U. Druskat, G. Mount, and F. Sala (Eds.), *Linking emotional intelligence and performance at work: Current research evidence with individuals and groups* (pp. 319). Mahwah, NJ: Lawrence Erlbaum Associates, Inc.

[37] Emmerling, R. J., and Cherniss, C. (2003). Emotional intelligence and the career choice process. *Journal of Career Assessment*, 11(2), pp. 153-167.

[38] Puffer, K. A. (2011). Emotional intelligence as a salient predictor for collegians' career decision making. *Journal of Career Assessment*, 19(2), pp. 130-150.

[39] Savickas, M. L., and Porfeli, E. J. (2012). Career adapt-abilities scale: Construction, reliability, and measurement equivalence across 13 countries. *Journal of Vocational Behavior*, 80, pp. 661-673.

[40] Douglass, R. P., and Duffy, R. D. (2015). Calling and career adaptability among undergraduate students. *Journal of Vocational Behavior*, 86, pp. 58-65.

[41] LaGree, D., Tefertiller, A. and Olsen, K. (2020). Preparing Mass Communications Students for an Evolving Industry: The Influence of Emotional Intelligence and Extracurricular Involvement on Career Adaptability. *Journalism & Mass Communication Educator*, pp. 1-13.

[42] Cooper, R. K. (1997, December). Applying emotional intelligence in the workplace. *Training & Development*, *51*(12),

[43] Goleman, D. (1995). Emotional intelligence: Why it can matter more than IQ. *Bantam Books*.

[44] Fosslien, L., and Duffy, M. W. (2019). No hard feelings: The secret power of embracing emotions at work. *Penguin*.

[45] Petrides, K. V., and Furnham, A. (2006). The role of trait emotional intelligence in a gender-specific model of organizational variables 1. *Journal of Applied Social Psychology*, 36(2), pp. 552-569.

[46] Tagiya, M., Sinha, S., & Chakrabarty, A. (2020). Emotional Intelligence of Mid-Level Service Manager on Career Success: An Exploratory Study. *Available at SSRN 3517672*.

[47] Spurk, D., Abele, A. E., & Volmer, J. (2011). The career satisfaction scale: Longitudinal measurement invariance and latent growth analysis. *Journal of Occupational and Organizational Psychology*, *84*(2), 315-326.

[48] Davies, K. A., Lane, A. M., Devonport, T. J., & Scott, J. A. (2010). Validity and reliability of a brief emotional intelligence scale (BEIS-10). *Journal of Individual Differences*.

[49] Scandura, T. A., & Ragins, B. R. (1993). The effects of sex and gender role orientation on mentorship in male-dominated occupations. *Journal of vocational behavior*, *43*(3), 251-265.

Challenges and opportunities of health sector entrepreneurship in Arunachal Pradesh

M Momocha Singh[1], Chaitan Kumar[2] and T Tadhe Goyang[3]

[1]Asst Professor, Department of Management & Humanities, NIT, Itanagar, Arunachal Pradesh 791112, India

[2,3] PhD Scholar, Department of Management & Humanities, NIT, Itanagar, Arunachal Pradesh 791112, India

Email:momocha@nitap.ac.in

Abstract. Arunachal Pradesh is a small state of eastern Himalaya, with 69 percent of the scheduled tribal population. On the good health and well-being, it has been ranked 23rd among 28 states. Seeing the constraints in the public health systems researcher reviewed the current health situation and key challenges where entrepreneurs can intervene. Secondary data and research reports were consulted for this paper. The indicators related to Non-communicable diseases and reproductive child health area are alarming. The rural area required more innovations and future solutions. Few diseases like the raise of Cancer cases will lead to further socio-economic burden. Health sector entrepreneurship is in the nascent stage due to a lack of human resources and remoteness. Digital intervention is very challenging due to the infrastructure shortage of mobile and internet services in the state. On the other hand, the state has a high potential of corporate social responsibility intervention as few big public sector companies are working in the state. New medical college and nursing intuitions will further supply the gap of future human resources. Herbal and other medicinal plants products related opportunity is huge in the state due to geographical advantages.

1. Introduction

The health sector's compound annual growth rate in India is 16.28 percent. It is expected that the market size will be 372 US $ Billion in 2022. Government expenditure is only 1.6 % of the GDP in 2020 (IBEF, 2020). Arunachal Pradesh with a population of 1383727 is situated in the eastern Himalaya region with a population density of 17 per square kilometer (Census, 2011). Recent data shows 78 % population has been enrolled in ADHAR (UIDAI, 2020).The majority population (69%) consists of schedule tribes. Overall SDGs ranking of the state is 26th and on the health sector, it has positioned at 23rd among 28 states. (NITI Aayog, SDG ranking, 2020). The literacy rate of the state is 65.4 % (Census, 2011). Health conditions in the Northeastern region are poor as compared with other parts of India. Arunachal Pradesh has been progressed a lot on various indicators. The state has adopted the public private partnership (PPP) model of primary health care services in 16 primary health centres which are being managed by non-profit organizations. The state has a limited number of private health sector entrepreneurs. The paper aimed to find out the future challenges and scope of health sector entrepreneurship in Arunachal Pradesh with the following objectives.

1. To analysis the current health situations of Arunachal Pradesh.
2. To find the future entrepreneurship opportunity based on the health sector gaps in the state.

2. Data sources and methodology

The paper used the data on secondary resources. Reports and policies of the government of India were reviewed. World health organization perspective and a few similar studies on the sector were consulted to find the overall health situation of the state. Census 2011 data and NITI Aayog data on Sustainable development Goals (SDGs) were reviewed. Key health indicators related to infrastructure, health establishments, human resources, disease burden, other health indicators' achievements were analyzed.

3. Key findings on health situations analysis of Arunachal Pradesh

Key indicators related to basic health like hospitals, beds, human resources, disease burdens, and achievement on health indicators of Arunachal Pradesh has been found al lower level among the various state of India. Health facilities in the state are of three categories namely primary, secondary and tertiary care centres. Key findings are summarized in following sub sections.

3.1 Health facilities

The state has 218 hospitals and 2404 beds as on as 31.12.2018 (NHF, 2019). The health centres have been categorized as in the table1.

Table 1. Health centres in Arunachal Pradesh.

Health centre type	Number of centres
Sub Centres	307
Sub Centres (Health & Wellness Centres) HWC	78
Primary Health Centres (PHC)	105
Primary Health Centres (PHC)-HWC	42
Community Health Centres (CHC)	63
District Hospitals	17
Medical Collage & Hospital	1

16 PHCs are being managed by public-private partnership (PPP) model in the state with the partnership of non-profit organizations.

The state has 3 corporate social responsibility (CSR) run primary health centres in their operation area which also caters general public along with company employees. 79 new Health & Wellness Centres (H&WCs) has been proposed in 2019-2020. Due to geographical difficulties health centres are being located in small population vicinity as compared to other states.

The state has adopted universal health coverage for the hospitalization of patients. An amount of Rs 500000 is the limit for each family per year for any hospitalization. The scheme has two components, central scheme of Prime Minister Jan Arogya Yojana (PM-JAY) and state scheme of Chief Minister's Arogya Arunachal Yojana (CMAY). Under the scheme, people can avail services outside the state in both public and private empanelled hospitals.

3.2 Human resources and capacity building

Total physicians, nurses, and midwives per 10000 population are 24 which are targeted 45 by 2030.Skilled human resources for health services in the state are detailed in the table2 (NHP, 2019).

Table 2. Skilled human resources.

Type of Professionals	Number	Reporting Year
Doctors	978	2018
ANM	971	2017
RN & RM	938	2017
LHV	15	2017
Pharmacist	279	2017

In the rural area the doctors- (129), health Assistant (87) health workers/nurses (573) were found shortage as 31.12.2018 (NHP 2019).

Capacity building of human resources is the key concern. The state has one new medical collage where first batch of doctors' degree education has been started. The state also started nursing schools recently for BSc nursing courses. For other Paramedical staffs training institutes has been established. Current staffs are being trained under various health programmes. Skilling and re-skilling of professionals is the demand. Community level health workers like Accredited Social Health Activist (ASHA) workers are being trained regularly on various community health modules.

Availability of skilled human resources and infrastructures can develop medical tourism in future. (FICCI, 2017).

3.3 Disease burden and risk factors among population.

State level diseases burden study shows that the state has premature death of 63.7 percent and disability or morbidity is 36.3 percent.Figure1 shows the ratio of morbidity and premature death in the population during 2016 (GDB, 2016).

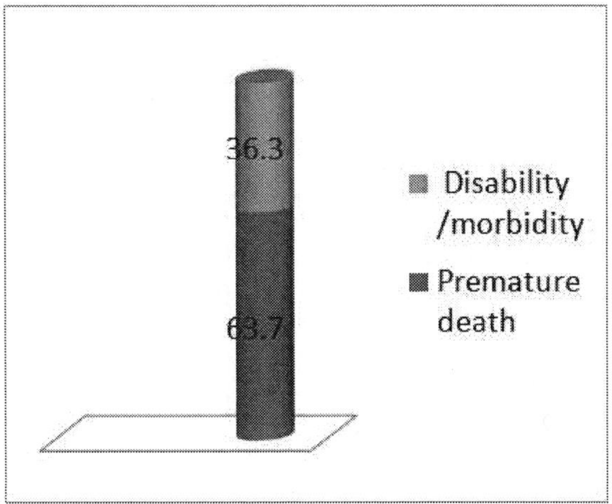

Figure 1. Death and Morbidity burden in Arunachal Pradesh.

Non communicable disease (NCD) burden is highest category which is approx 53 % and communicable, maternal, neonatal and nutritional diseases (CMNNDs) are second cause with 35.5 % burden and injuries related burden is 11.6 during 2016. Figure 2 describes the ratio of diseases burden among three categories. Here NCDs are mainly life style diseases.

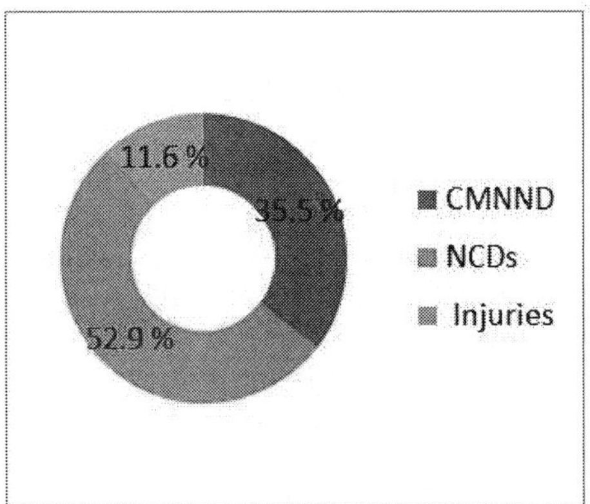

Figure 2. Percentage of diseases categories.

Iron deficient anaemia was found topmost cause of deaths (14.8 %) in the study.

Future expected infection of Cancer in the population ranges from 1 in 4 to 1 in 8 which is highest among various states (ICMR, 2020). Annually more than 1200 cancer cases reported and more than 600 deaths due to cancer reported in the state during 2014 to 2016 (ICMR,2020).

45 % of people consume tobacco in various forms (GATS 2).The Tuberculosis notification rate per 100000 populations is 216 which is higher than the national average of 160.

Screening of NCDs at public sector health facility is approx 7000 for 18 month period during April 2018-September 2019. Here more demand can be created through innovative entrepreneurship approaches.

Top ten risk factors which contribute to 50.9 % of overall death and disability in the population are listed in the table 3 (GDBS, 2016)

Table 3. Top ten risk factors of death and disability.

Percentage	Contribution RiskFactor
Malnutrition (Child & Maternal)	14.8%
Air Pollution	5.1%
High Blood Pressure	5.0%
Tobacco Use	4.9%
Dietary risks	4.7%
High fasting plasma glucose	4.0%
WaSH (unsafe water, Sanitation and hand washing)	3.9%
Alcohol & drug use	3.9%
Occupational risks	2.6%
Impaired kidney function	2.0%

3.4 Health indicators

The state ranked 23rd in overall health indicators in the 2019-2020 ranking. As per SDGs indicators by NITI Aayog the state needs a high push on the following indicators.

The Infant mortality rate is 36 (SRS 2019).The pregnant women aged 15-49 years who are anemic are 33.8. Children aged 6-59 months who are anemic is 28.3 %.The Proportion of institutional deliveries is 45.3 %.The Under-five mortality rate per 1000 live births is 33. Full immunization for the 0-5 age of children is 41.3 %. The Institutional delivery is 45.3%.

As per the health management information system (HMIS) report mothers who had at least 3 Ante-Natal care checkups is 30.5 percent (HMIS 2018-19).

Above key health indicators required high focus and entrepreneurs can bring innovative models for affordable solution.

3.5 Platform for social entrepreneurs in heath sector

WHO advocated social entrepreneurship is the health sector for affordable quality services (WHO, 2006). Health sector entrepreneurship is considered as social entrepreneurship where innovative solutions are required for such hard to reach area and also to bring the cost of services lower without compromising the service quality. National health policy 2017 also advocates partnership approach for bringing affordable solutions. The state government start-up policy has provision for incentives and incubation support for health sector start-up.

R K Mission Hospital state-granted non-non profit private hospital is the best example for social entrepreneurship in the state which serves tertiary care services in the state.

Private hospital-like Heema Hospital is another example of private sector secondary and tertiary care services in the state.

The state has Corporate Social Responsibility (CSR) resources. National Hydropower Corporation (NHPC) and North East Power Corporation (NEPCO) are the biggest public sector corporations are complementing the health facilities with their CSR capacities.

Few districts including capital complex have few successful Private clinics that cater to the services. The rural area is highly underserved due to various resource scarcity and patients are visiting outside the state to avail of medical services.

The state has opened a new medical college for doctor's educations. Nursing schools both in private and public mode are functional which will bridge the gap of human resources in health sectors and future entrepreneurship opportunities will be opened. Skill development institute is also supported by the national livelihood mission programme for short term courses for catering to the demand for skilled manpower.

The state has adopted a clinical establishment act that facilities the private sector to venture into the sector.

3.6 Key Challenges for the entrepreneurs:

There is difficulty in transportation due to geographical terrain challenges. Lower population density, is another business challenge. Public infrastructures like road connectivity; regular supply of electricity is the key for health infrastructure building.

In the state mobile tele-density is 7.42 % and Internet subscribers per is 25.88 % in the population. Availability of internet services in the remote area is very poor. This was found the biggest challenge for digital health coverage in the state.

Availability of trained human resources for running such enterprises will be another challenge in the state. Capacity building itself can be one of the opportunities for the entrepreneurs.

Health service seeking behaviors of the public to be studies before venturing into any area in the state.Ongoing infrastructure development for transportation and digital communications will further provide base for entrepreneurs.

4. Opportunities for health sector entrepreneurship

A household using clean cooking fuel is 64.7 %. Clean fuel and other sanitation based entrepreneurship can be future options.

Tobacco consumers in the state is more than 45 % .Alcohol consumption is also very high. Social entrepreneurs can venture on this area for health prevention activities.

Herbal and medicinal plant cultivation and processing businesses have vast opportunities in the state due to geographical advantages. This will also help for Immunity Busting in the population. More than 500 medical plants are found to use in the state.(Tonlong and Sumpam, 2020).

Telemedicine services like the public sector "eSanjeevani' and many private players have been growing rapidly in India. The state has a huge scope of telemedicine service in the future.

Government of India has provided Telemedicine guideline 2020 and announce national digital health mission for future digital health adoption in India which will support the entrepreneurs on health sector immensely.

Few future opportunities have been summarized in Table 4.

Table 4. Future Business Opportunity.

Future Business Opportunity	Focus area	Opportunity Factor
Preventive health check up / Screening services	Non communicable diseases (NCDs)	High burden of NCDs/Cancer
Tobacco /alcohol session	Counselling / Telemedicine	More than 45 % consumers.
Anaemia prevention	Detection& Prevention	Diagnosis, Nutrition
Immunization	Vaccination	Follow up

strengthening	outreach	and delivery
Herbal products manufacturing	Aromatic, medicinal plants products	Availability of more than 500 species.
Digital health cares	Telemedicine's &Digital health	Future, Innovative in the hilly area

5. Conclusion

The state has found legging behind on various fronts of health indicators. Innovative solutions required to meet the challenges of physical Infrastructure, shortage of trained human resources. Technological innovations can create value in the area of the gaps. Telemedicine and digital health will further help to boost the state if the mobile and internet services are available adequately. The state can incentivize the health sector investors. Entrepreneurs who are willing to work in the high burden area of health sector require further detailed research on concern area. Return on the investment will be key concerns as investment in the health sector gives long term returns. Hence government incentives are required for the entrepreneurs to innovate and implement. The State may have provision for research and development for innovation solution. Universal coverage for preventive health care services is highly demanded so that in future costs on secondary and tertiary care will be reduced. Social entrepreneurs can join hands with the state to bridge the gaps. The state has technical institutions like North Eastern Regional Institute of Science and Technology (NERIST), National Institute of Technology (NIT) and Tomo Riba Institute of Health And Medical Sciences (TRIHMS) who can play leading role along with other public and private universities. All can join hands for start up development on the health sector. Affordable, accessible health services along with public and private sector collaboration can be achieved. Health sector entrepreneurship is required to scale up of services and achieving sustainable development goals of the state. Social enterprises in the health care sectors can further fill the gap of services and sustainable development goals can be achieved on time.

Appendix: Glossary

Words	Expansion
IBEF	India Brand Equity Fund
UIDAI	Unique Identification Authority of India
SDGs	Sustainable Development Goals
NHP	National Heath Profile
GDBS	Global Diseases Burden Study
ICMR	Indian Council of Medical Research
GATS 2	Global Adult Tobacco Survey
SRS	Sample Registration System
HMIS	Health management Information System
WHO	World Health Organization
NCDs	Non communicable Diseases

References

[1] India Brand Equity fund (2020). *Growth of Healthcare Industry in India* – Infographic, (https://www.ibef.org/industry/healthcare-india/infographic).

[2] Government of India (2011*).Census of India* 2011

[3] Unique Identification authority of India, (2020). State wise Aadhar saturation, (https://uidai.gov.in/images/state-wise-aadhaar-saturation.pdf, updated on 31 May 2020)

[4] National Institution of Transforming India (2020). *SDGs India Index and Dashboard* 2019-20, (https://niti.gov.in/sites/default/files/SDG-India-Index-2.0_27-Dec.pdf)

[5] Ministry of Health Family Welfare, Government of India (2019). *National Health Profile*,2019.

[6] FICCI (2017). Developing service sectors Linkage in North East

[7] Global Burden of Disease Study (2016).*The burden of cancers and their variations across the states of India: the Global Burden of Disease Study* 1990–2016, (https://www.thelancet.com/journals/lanonc/article/PIIS1470-2045(18)30447-9/fulltext)

[8] Indian Council of medical research (2020) .*Report of National Cancer Registry Programme* 2012-2016, (https://www.ncdirindia.org/All_Reports/Report_2020/default.aspx)

[9] Indian Council of Medical Research (2016) .*A Report on Cancer Burden in North East India,* 2012-14.

[10] Global Adult Tobacco Survey (2017).*India Fact Sheet* 2016-17.

[11] Press information Bureau, Government of India (2020).*Screening of Non Communicable diseases*, March 2020.

[12] National Institution of Transforming India (2020). *SDGs India Index and Dashboard* 2019-20.

[13] Sample Registration System, Government of India (2019).*SRS Bulletine*, 2019.

[14] Heath Management Information System, (2020). *Arunachal Fact Sheet,*HIMS, (http://www.rrcnes.gov.in/hmis/2019-20/arunachal_factsheet_19-20.pdf)

[15] World Health Organization (2006).Integrating *social entrepreneurs into the "health for all" formula* (http://www.who.int/bulletin/volumes/84/8/06-033928/en/).

[16] Government of India (2017).*National Health Policy-2017.*

[17] Press Information bureau , Government of India (2020).*A big win for Digital India: Health Ministry's 'eSanjeevani' telemedicine service records 2 lakh tele-consultations on 19 Aug 2020,* (https://pib.gov.in/PressReleseDetail.aspx?PRID=1646913)

[18] Tonlong W. and Sumpam T. (2020). *Status of Medicinal Plants in Context of Arunachal Pradesh,*(https://www.springerprofessional.de/en/status-of-medicinal-plants-in-context-of-arunachal-pradesh/16890006)

[19] Ministry of Health Family Welfare, Govermnet of India (2020),*Telemedicine Practice Guidelines, March 2020*, (https://www.mohfw.gov.in/pdf/Telemedicine.pdf).

CONTRIBUTION OF DEPARTMENTAL STORES TOWARDS SOCIO-ECONOMIC TRANSFORMATION IN PAPUMPARE DISTRICT OF ARUNACHAL PRADESH

Miss Laxmi Rai[1]
Reasearch scholar,CMS
North eastern regional institute of science and technology ,nirjuli
Itanagar -791111, Arunachal Pradesh. Email : laxmirai605@gmail.com

Dr Manmohan Mall[2]
Assistant professor
Centre for management studies NERIST, Nirjuli
Itanagar -791111, Arunachal Pradesh. Email: Mallmanmohan79@gmail.com

---

Abstract

Retail sector has been booming in India with its share of around 10% of the country's Gross Domestic Product and that absorbs around 8% human resource. The experiment of Departmental Stores was initiated with the view to offer various diversified goods and services under one umbrella. A school of thought believes the concept has been developed from the rural "Haat' where a particular market place is chosen where all the vendors come to that place with their goods on a prefixed day of a week. However, the Departmental Stores primarily deal with branded goods and it is mostly concentrated in the cities. Of late, the penetration of Departmental Stores has been profoundly witnessed in the urban areas and even in some of the district headquarters or big towns. It is interesting to know how the sector has been performing in the three-tier cities or towns. In Arunachal, the similar wave of establishing Departmental Stores is being witnessed particularly in the twin cities of the state. There are National Stores as well as Local Stores that have been operating since 2014. The Capital Complex of the state has a total number of 12 stores. This paper has attempted to understand the status and its ease of contribution towards socio-economic life of the city dwellers.

Key Words: Departmental Stores Socio-Economic Retai,Haat And National Stores

---

1. Introduction

Departmental Stores ensure to make variety of products available to customers. Apart from that, in order to attract customers, they also provide various facilities including credit, carrying goods to their and other services as well. Besides, it also improves the standard of living of the people by providing employment opportunity in quite a large scale.

In Arunchal Pradesh also, Departmental Stores contribute immensely towards the development of the economy. The sector has generated considerable amount of employment for the citizen of the state over a period of time. Therefore, this chapter presents the glimpse of its contribution towards socio-economic transformation in Arunachal Pradesh in general and Capital Complex, Itanagar in particular.

1.1 Research Methodology

Sampling Technique A multi-stage sampling technique has been used to arrive at final selection of samples for the present study. **Data collection:** For the present study, both primary and secondary data were collected. Primary data were collected from the study area by using structured questionnaire whereas secondary information were gathered from the sources like published research articles, published theses or dissertations, internet, magazines, newspapers, etc. **Universe/Population of the Study:** Twelve (12) Departmental Stores namely, V2, Reliance Trends, Vishal Mega Mart, Bazaar India, InMark, M-mart, Sohum, City-Arcade, V-Mart, Pantaloons, NE Departmental Store and LK-Hub were the universe of the study. **Total Selected Samples:** Out of twelve (12) Departmental Stores mentioned above, only eight (08) Departmental Stores viz., M-mart, Sohum, V2, Reliance Trends, Vishal Mega Mart, Bazaar India, LK-Hub and NE Departmental Store have been selected for the present study on the basis of the criteria determined by the researchers. **Tools used for data collection:** Structured questionnaires were used for the purpose of data collection from the study region. **Data Analysis:** Data analysis has been done through statistical tools like SPSS, Excel etc. After analysing data, the results were interpreted accordingly and reached appropriate conclusions. **Periods considered for the study:** Three years i.e., 2017, 2018 and 2019 have been considered as the study period.

1.2 Limitations of the Study

Following are some of the limitations of the present study-

- The study is very small as it has only eight samples due to which the results of the present study cannot be generalized.

- The researchers had to squeeze the study to a great extent as the time was very limited. As such, very limited socio-economic variables have been considered which may not be justified.
- Questionnaire was prepared only for Departmental Stores whereas the results would have been more realistic and practical, had customers were also interviewed through well structured questionnaire.

1.3 Literature review

Goswami, K. & Goswami, G. (2017) carried out study on "An Analysis of Growth of Fashion Retailing in Guwahati City" about the growth rate of the apparel and accessories market in the Guwahati City. It was found that Fashion Retailing has huge market potential as the size of population in the city is also quite large.

Ubeja, S.K. & Bedia, D.D. (2011) had undertaken a study on "A study on customer satisfaction towards Departmental Stores in Tripura district". The objectives of the study were to analyze the customer satisfaction with respect to products, price, place and promotion in Indore city and to study the variation in customer situation due to different demographic variables. The result reveals that satisfaction level of the customer toward the six factors extracted through factor analysis is significantly influenced by the socio-economic factors.

Das, G., Guin, K.K. & Datta, B. (2013) in their research articles "Impact of store personality antecedent on store personality dimension: An empirical study of department retail brands" tried to explore the impact of store personality antecedents on store personality dimension. The survey was done by using systematic sampling of department store shoppers of age 18 years and above in Kolkata. They found that the different sets of store personality antecedent affect various Departmental Storepersonality dimensions differently.

Panda, A. (2013) has conducted a study on the topic "Customer Patronage towards Food and Grocery Retail''. The objective was to find out the decision variables which influence the selection of a traditional outlet or a modern store for the purchase of food and grocery item and found that the traditional outlet or formats more beneficial than the modern format related to convince services and easy policies are available in term of exchange and return of goods.

Kaur, P. & Singh, R. (2007) studied on the topic "Uncovering Retail Shopping motives of Indian Youth" The objectives of the study was to bring a huge opportunities for foreign as well as domestic players in India in the field of retail. This study was done with the help of a sample of 115 students and their response had been measured through a personal administer, structured questionnaire. The study reveals that the Indian youth usually shop with a self- obsessed perspectives.

Many literatures are available with regard to the study of Departmental Stores. However, no literature exists regarding Departmental Stores in Arunachal Pradesh. That is where the research gap lies as no study has been undertaken on Departmental Stores in Arunachal Pradesh so far. Taking this gap into consideration, an attempt has been made to study the Departmental Stores in Arunachal Pradesh.

2. Analysis and Interpretation

The primary data were collected from the field using appropriate questionnaire and the same has been properly analysed and interpreted as follow-

Table 1: Total Number of Employees: Year-Wise

YEAR / STORE	2017	2018	2019	Grand Total
National Stores	319	342	354	
Local Stores	51	68	73	
Total	**370**	**410**	**427**	**1207**

The above table depicts the employment generation by both National and Local Departmental Stores. In case of National Stores, there is an increasing employment trend from the year 2017 to 2019. In the year 2017, 319 employments have been generated followed by a significant increase in the employment rate in the year 2018 which stands at 342. The increasing trend of employment persists in the year 2019 as well; however, the employment trend is increasing in a diminishing rate.

On the other hand, Local Stores also display a decent growth in terms of employment generation. In the year 2017, 51 employments have been generated followed by 68 in the year 2018 and 73 in 2019. Therefore, the dataset shows an increasing trend of employment by Local Stores.

It is fascinating to see that the employment trend in both National and Local Stores are significantly increasing. However, National Stores are contributing more in terms of employment generation as compared to Local Stores.

It is also worth mentioning that a massive employment of 1207 has been observed which can be considered as a positive indicator for the socio-economic upliftment of the indigenous people in particular and the state in general.

Table 2: Gender-Wise Employment Distribution

YEAR/GENDER STORE	2017		2018		2019		TOTAL	
	Male	Female	Male	Female	Male	Female	Male	Female
National Stores	141	178	149	193	154	200	444	571
Local Stores	20	31	32	36	37	36	89	103
Total Male & Female employees							**533**	**674**

The above dataset shows that in both the stores, female has dominance over male counterpart in terms of employment. Therefore, the Departmental Stores are making positive contribution towards Gender Equality which is one of the most significant goals among 17 SDGs (Sustainable Development Goals) of United Nations.

Employment trend of National Stores is ever increasing in case of both male and female whereas female employees are more than male employees which shows that there is extensive opportunity for female population in retail sector. Similarly, Local Stores are also contributing on their part towards female employment generation but in small scale as compared to National Stores.

Table 3: Local & Non-Local-Wise Employment Distribution

YEAR STORE	2017		2018		2019		TOTAL	
	Local	Non-Local	Local	Non-Local	Local	Non-Local	Local	Non-Local
National Stores	217	102	216	126	226	128	659	356
Local Stores	38	13	48	20	53	20	139	53
Total Local & Non-Local employees							**798**	**409**

As per the above table, it can be observed that huge employment opportunity exists in Departmental Stores operating in Capital Complex, Itanagar. In every year, employment generation for local people is more than that of Non-local people. The finding shows that the total employment of local workforce from last three years is 798 which are almost double to the total employment of non-local workforce i.e., 409.

Weighted Average Salary of Employees Working at the Bottom Level of the Selected Departmental Stores in Capital Complex, Itanagar, Arunachal Pradesh

Weighted average salary of employees

$$= \frac{7,500x44+5,000x75+7,500x34+12,500x68+7,500x58+7,500x26+5,000x10+5,000x56}{371}$$

= 27,70,000/371

= Rs. 7,466.30 (*Rs.6,750 per month as per 7th Pay)*

This shows the single digit (i.e., 8) Departmental Stores can accommodate more than 1200 employees of which 371 employees belong to Bottom/Lower Level and most significantly, the weighted average salary is around 7,500 per month augmented with other incentives that may be further augmented by incentives and bonus. The dataset exhibits that the participation of women employees are always higher than the male in every level and in almost all Departmental Stores which essentially propelling our commitments towards achieving Sustainable Development Goals (SDGs) and Gender Equality (SDG number : 5) in particular. The state has its Geo-political strategic position which solely depends on household farming and fewer government jobs as the state is deprived of big private investments. The Departmental Stores have been coming-up with its multiplier benefits to the people of the state that includes distribution of goods, availability of variety of products/brands and of course, potential opportunity of employment generation.

3. Findings

Through the present study, researchers have come across the following findings-

1. The study reflects that the Departmental Stores are growing in the region since 2014. However, in Nirjuli and Doimukh market, there is no such facility available.
2. The study indicates that the sector has been absorbing growing number of manpower and it has the potential capacity to grow faster since the region is required to further explore.
3. These organized retail majors are committed to gender equality rather the sector has absorbed more number of women than its counterpart.
4. Although a few firms may not have satisfactory growth story but it essentially needs little more gestation period since the region is emerging but still lagging behind the mainstream economy.
5. The results reveal that the sector is promoting the indigenous people that are significant for achieving rapid socio-economic transformation.

4. Recommendations

Following recommendations are made by the researches in the line of above findings-

1. Initiative must be taken to expand Local Brands so that untapped resources of the state can be marketed utilising the experiences and distribution channels of the giant retailers.
2. The Local Brands or manufacturers can get the access to market their product through these major Departmental Stores/Chains so that appropriate focus may be given on the Consolidation of Local Brands to make them competitive with the National Brands.
3. It may ensure the Creation of Women Entrepreneurs through SHGs who can transform the unexplored local based resources mainly the agro-based resources to make some innovative products.
4. Budding entrepreneurs must be motivated to come forward and the government must train and groom them through various Training and Workshops so that they can start their own ventures in this field. The indigenous firms may be mentored and supported by appropriate nodal agencies so that they can also operate in other places by crossing the state boundaries.
5. If the retail chains of these giant firms can be utilized, it would definitely open the flood gate of promoting and incentivizing local and unique products without creating a separate marketing infrastructure.

5. Conclusion

From the above results, it is quite obvious that Departmental Stores have the power to transform socio-economic condition of the people of Arunchal Pradesh in particular and country in general. It is evident that Departmental Stores have created considerable amount of employment for the state in last few years. In fact, it provides more employment opportunity to indigenous/local people as compared to Non-Locals. Most interestingly, it creates a very good platform for women as the dataset shows that more women are employed as compared to male counterpart in both National and Local Departmental Stores. Therefore, it can help the state in achieving UNs' SDGs like Gender Equality, Unemployment and Poverty which are the global concern.

References:-

1. Bitner, M. J. (1990). Evaluating service encounters: the effects of physical surroundings and employee responses. *Journal of marketing*, *54*(2), 69-82.
2. Das, G., Guin, K. K., & Datta, B. (2013). Impact of store personality antecedents on store personality dimensions: An empirical study of department retail brands. *Global Business Review*, *14*(3), 471-486.
3. Gomati, R., Kishore, S., & Deepika, R. (2013). A Study on Customer Satisfaction towards Departmental Stores in Erode City. *International Journal of Engineering and Management Research (IJEMR)*, *3*(5), 38-41.
4. Goswami, K., & Goswami, G. (2017). An Analysis of growth of Fashion Retailing in Guwahati city. *Journal of Marketing Vistas*, *7*(1), 12-18.
5. Grönroos, C. (1984). A service quality model and its marketing implications. *European Journal of marketing*, *18*(4), 36-44.
6. Kaur, P., & Singh, R. (2007). Uncovering retail shopping motives of Indian youth. *Young Consumers*, *8*(2), 128-138.
7. Manjula, K. (2017). A study on customers satisfaction towards departmental stores with special reference cuddalore region. *Journal Homepage: http://www. ijmra. us*, *7*(10).
8. Oyeniyi, O. J., Abiodun, A. J., & Osibanjo, A. O. (2012). Job Satisfaction and Intention to Leave among Nigerian Salesmen. *Journal of Management Research*, *4*(2), 303-314.
9. Oyeniyi, O., & Abiodun, A. J. (2012). Measuring retail service quality in Nigerian departmental stores. *Journal of Economic Behavior*, *2*, 37-45.
10. Panda, A. (2013). Customer Patronage towards food and grocery retail-a case study. *Global Journal of Management and Business Studies*, *3*(9), 955-960.
11. Rajaram, P., & Sriram, M. V. (2014). Measuring retail service quality: A study on indian departmental stores. *International Journal of Business Management & Economic Research*, *5*(1).
12. Ramayah, T., & Jasmine, Y. A. L. (2003). What drives relationship quality? A study on two retail clothing stores, The Proceedings of the 5th Asian Academy of Management Conference, Challenges of Globalized Business: The Asian Perspective.
13. Ramya, N., Ali, S. M., & Bhuvaneshwari, A. (2017). A study of consumer buying behaviour in departmental store (with special reference to Coimbatore city). *IJAR*, *3*(3), 734-737.
14. Siu, N. and Tak-Hing Cheung, J. (2001), "A measure of retail service quality", *Marketing Intelligence & Planning*, Vol. 19 No. 2, pp. 88-96.
15. Ubeja, S. K., & Bedia, D. D. (2012). Customer satisfaction in shopping malls: an empirical study. *Pacific Business Review International*, *5*(2), 60-72.

Published Thesis

1. Sen, S. (2013). Design and optimization of vision based light control system for smart Departmental Store. Dept. of Electrical & Electronic Engineering, BRAC University, Bangladesh.

Web Sources

1. Retrieved from http://www.pondiuni.edu.in/storage/dde/downloads/markiii_rm.pdf, accessed on 27.09.2019
2. Retrieved from https://www.thebalancesmb.com/what-is-retail-2892238, accessed on 30.09.2019.

3. Retrieved from http://www.yourarticlelibrary.com/stores/retail-stores-store-based-retailer-and-non-store-retailing/7586, accessed on 10.10.2019.
4. Retrieved from https://en.wikipedia.org/wiki/Department_store, accessed on 16.10.2019.
5. Retrievedfrom https://shodhganga.inflibnet.ac.in/bitstream/10603/102750/10/10_chapter%203.pdf, accessed on 15.10.2019.
6. Retrieved from http://indianresearchjournals.com/pdf/IJMFSMR/2012/July/16.pdf, accessed on 20.10.2019.

A Conceptual Framework of E-Shopping Behaviour of Millennials in the Present Era- An Indian Perspective

Priyanka Chetia[a], Dr. Manmohan Mall[b]

[a&b]NERIST, Nirjuli, Arunachal Pradesh, Pin: 791109

poo.chetia@gmail.com

Abstract. The present era is the digitalization era and the advanced technology act as an emerging drive for many changes and such changes have a direct impact on our daily lives. Technology thus impact the nature of work and also lead towards the growth of a latest generation of millennials who are also called as "Generation Me","Digital Natives" and "Generation Y". It has become a tough job in the part of the retailers to acquire these new generation customers and to keep the old customers. Millennials preferably shop by taking into mind the comfort and availability of the product they want. Thus, it is necessary for the retailers to understand their customers and their choice of shopping to get the competitive advantage over the available shopping channels. The population who are using mobile internet is new. Even though there was the use of mobile phones in the past few years but the new population is more technologically advanced, mobile enabled and are thereby connected more socially. The growing internet users have paved the way to online retailing. In this paper we study the millennials who prefer to spend money on online shopping than any other shopping channels. This paper reviewed the studies of millennial shoppers and their shopping trend in India. This will help the retailers to understand the issues and doubts related to the buying process of the millennial customers. Thus this paper is an attempt to examine the Indian millennial buying habits and is thereby helpful for the companies who are trying to acquire the millennial consumers.

1. Introduction

In the recent couple of years we have been observed a new retail experience. This changed experience is due to the change in technology. The technology has made everything more mobile and the virtual interface and gesture based technology have renewed retailing into a totally different level. The availability of shopping channels has made it difficult for today's consumers to choose the mode of shopping and to get their satisfaction. Today's empowered customers prefer to shop with the comfort and their convenience of shopping. Nobody wants to go beyond their comfort zone and make their purchases at their convenience level. According to the report of ASSOCHAM, 2013, more than 50,000 crores size of business has come from Indian retail market. Moreover an annual 100 percent growth rate has been forecasted in the same report. The trend of E-shopping has witnessed an upward move in India. The use of fast growing technology and huge number of web users in India, there is a thriving scope for numerous e-retailers to expand their business. India has become an international hub of online business. Earlier, it was said that "change" is the only permanent thing in the world but in today's world, the rate of "change" has been changing rapidly and the changing time has also arrived in India. It has become very important to understand the demographic profile of the customers in order to meet their needs. The high speed mobile internet connectivity, higher disposable incomes, changing lifestyles has made the online channels to become more important among the millennials.

The availability of large quantity of choices for the consumers to purchase a product has the retailers have to face a serious challenge to maintain competitive advantage. Some customers go for online purchase while some other customers go for offline and many go for both online and offline purchase. Thus, it is very important for all the customers to decide the particular mode of shopping among the available shopping channels. The study focuses on the choice to shop among the Indian millennials. This paper is an attempt to study the shopping habits of the millennials which includes their preferences of choosing the shopping channels and choice of buying products so as to help the retailers to adopt new tools and techniques to attract more millennial customers.

2. Method

This study is based on the- review of various past literatures related to the millennials and a conceptual study was conducted through the review of work done in the area of young generations and their buying behavior. This analysis revealed the present facts about millennials and their buying behavior and its impact on e-shopping. It is a qualitative research with secondary data analysis from various scientific articles including Taylor & Francis, Springer, Emerald, Elsevier and related websites etc.

From the past related literatures 41 articles were selected as primary corpus from journals which includes Journal of Retailing and Consumer Services (2014), International journal of Science, Technology & management (2015), Canadian Social Science (2015), International Journal of Engineering and Management Research (2016), International Journal of Advanced Research (2016), International journal of marketing studies (2017), Indian Journal of Industrial Relations (2017), Employee Relations(2018), Journal of Cleaner Production (2018), Academy of Marketing Studies Journal (2018), Journal of Business Research (2019), International Business Review (2019), International Journal of Applied Business and Economic Research (2019).

3. Literature Review

3.1. Millennials

The millennial Generation is the offspring from the boomers. They are also known as Generation Y. The millennial comprises of people who are born in between 1977-2000 (Noble, Haytko & Phillips, 2009). But in 2012, Bednall, Valos, Adam & McLeod shortened this interval to 1980-1994. During sixties, the different ways in the lifestyle and belief eco system reflect the age gap among the youngsters and their parents. The characteristics among the Millennials are measured on different parameters like the education, value system and buying behaviour. The appraisal highlighted beyond socio economic parameters (Guven Ordun, 2015). The millennials (born in between 1980-2000) are highly demanding and their expectations are also at the top of the mount. They are the trend- setters and are truly the digital consumers.

The millennials are also known as Generation Y, Nexters and Echo Boomers. The marketers are interested towards the Generation Y consisting of 56 million individuals (R.S. Sethi et. al, Nov.2017). The size of Generation Y is three times more than the size of Generation X. The Generation Y who is also called Digital Natives are grown up in the world of digital evolution. They are familiar with the basics of online marketing and most importantly they know their value as consumers (Emma Miller). Generation Z are those consumers who born after the year 2000. Firms are interested for the millennials who are more independent as compared to Generation Z as of today. In this digitalisation era, it becomes best to tap this new growing segment that is yet to explore especially with digital marketing strengthening its roots in India (Reetika Jain, 2018).

Table 1:

Millennials and their period of Birth

Name of the Author/Year	Birth Period of the Millennials
Valentine and Powers (2013)	1977-1996
Muda,Mohd, & Hassan(2016)	1980's to the early 1990's
Omar (2016)	1980-1990's
Lissitsa & Kol (2016)	1980-1999
Moore (2012)	1982-2000
Rainer & Rainer (2011)	1980-2000
Lee & Kotler (2016)	1980-2000
Junker,Walcher,& Blazek (2016)	1981-1995
Ordun (2015)	1981-2000
Howe & Strauss (2000)	1982-1988
Pasi Pyoria, Satu Ojala, Tiina Saar and Katri Maria Jarvinen (2017)	In or after the 1980s
Darija Aleksic and Kaja rangus(2020)	early 1980's and the late1990's

Source: Original writing from the author

3.1.1. Millennials and the Rural market in India

In India, it was estimated to increase the number of internet users in both urban as well as rural regions due to a dynamic growth in internet access. The internet users are expected to increase from 432 million in 2016 to 647 million by 2021, i.e.; the internet penetration from 30 percent in 2016 to 59 percent in 2021. Among them around 75 percent new internet users are expected to come from the rural area of the country (KONNECTED to consumers, Deloitte,2017).As per the census data, out of the total millennial population of India i.e; around 450 million, 300 million millennials (67% of total population) live in rural India. This 300million rural millennials represents 36 percent of the rural population which represent a significant market. Moreover, millennials monthly households income accounts for 80 percent on an average. This estimates a spend potential of USD 220 billion approximately which is influenced by the rural millennials (Kantar IMRB STAR report 2017). We have seen a growing digital footprint of the rural millennials which enables the degree of homogenization of expectations across the urban and rural divide in India, but there is a continuous control by the traditional media like TV and outdoors along with retail outlet, newspaper as a primary source of informations. Further, the study showed that the celebrities also influence the states and the impact on millennials is more in states like Maharashtra, Bihar and West Bengal and less impact in the northern states.

3.1.2. Characteristics of Indian Millennial Consumers

To understand how the millennial consumers think about a particular product and what they actually need from it is an important factor. One cannot generalize the characteristics of the millennials across the globe. In different geographical locations, the influencing factors like economics, politics and culture/social life style are also different. Taylor (2006) in his paper stated that in order to know the buying habits of the millennials, one must understand the communication methods and interaction of the millennials. Millennials born and grew amid the internet and technology and so are very accustomed with its uses. Use of mobile phones and other electronic gadgets become the life lines of the people and without them the panic situation will arise (Camarda, 2016). Thus internet and high use of smartphones and specially the active users on social media make this generation all over the world have the same cultural icons and life style trends. Hershatter and Epstein (2010) stated that millennials have been grown up in an environment where they get treatments like feedback, atention, praise and guidance for any activities performed by them and so the same thing they expect in their working environment too. Millennials are also characterised as individualistic, self-centered, optimistic, entitled and disloyal. Unlike the previous generation the millennials have higher levels of self esteem and they believe in themselves and control over things by their own (Darija et al., 2020).
Thus to understand the buying behavior of the new age generation, we must know their characteristics. According to the latest Deloitte India survey in 2019, the new age consumers have the following characteristics:

- The growing internet penetration has led to a increasing number of online shoppers in the country. Data showed that the m wallet transactions in India has an increase from INR 200 billion in FY16 to INR 3,000 billion in FY 18. Hence the purchase pattern of the consumers is changing.

- With the use of advanced technology, today's consumers are connected globally than ever before. The empowered consumers can get whatever they want, when they want and where they want.

- The savvy consumers of the new generation consider convenience as an important factor in case of purchasing and any other activities related to them.

- Shopping is not limited to shopping only for the new age consumers. They want to explore new things and a good shopping experience is becoming a key factor to choose among the shopping channels. Experiential shopping has taken place in the country.

- The consumers became more health conscious than before and they prefer more organic products including herbal and Ayurvedic.

3.2. E-Shopping and the Millennials

According to Statista 2020, around 54 percent of indian population are internet users who belongs to the age group of (20-39 years) in early 2019. Moreover, the number of internet users in India is projected to grow from 700 million users to over 974 million users by 2025. I-n 2019, Malhotra et al., in 2017 stated that e-loyalty is the key factor and it decides the- online business profit. Cyr, 2008 in his paper defined e-loyalty as the intention to revisit and repurchase from an e-commerce portal even when other options are also available. Technology Acceptance Model (TAM) has been used to determine the e-loyalty in extant studies by Davis in 1989, Chen et al., 2002, Cyr et al., 2006, Lu et al.,2003, Pantano and Priporas,2016,Venkatesh and Davis,2000. There are a number of factors that affect the consumers' willingness to shop online. Jozef et al., (2018) developed a structural model of seven factors based on the Purchase behaviour hierarchical model. These seven fators include the factors of price, availability, social proof, scarcity, product details, conditions and social media activity. A study was conducted on the college going studnets in India by Vaidehi (2014) and it was found that availability, low price, promotions, comparison, convenience, customer service, perceived ease of use, attitude, time consciousness, trust and seeking for variety are the influencing factors of e-shopping among them. Bhattacharya in 2001 stated that the ease of use and usefulness are the main factors in TAM models which determine the technology adoption and thereby determine online loyalty towards new tehnology. K. Purani et al., in their study e-xamines the indirect influence of personal characteristics like perceived compatibility, inovativeness and peer influence in terms of perceived subjective norms. Their findings suggest that while deciding the e-loyalty of the millennials, personal characteristics and social influence play the Avital role. But when it comes to generational cohorts like the millennials, only the technical or relational characteristics may not define their loyal behavior. As millennials grew with and amid of technology unlike the previous generation and thus the impact of technology has on their cognitive, affective and social outcomes (Immordino-Yang et al., 2012). Determinants of online information search model showing the factors includes ease of use of technology, effectiveness of online search and user satisfaction are antecedents of the perceived benefits of information search. Each of these factors are shown in Figure:1 (Kulviwat et al., 2004)

Figure:1

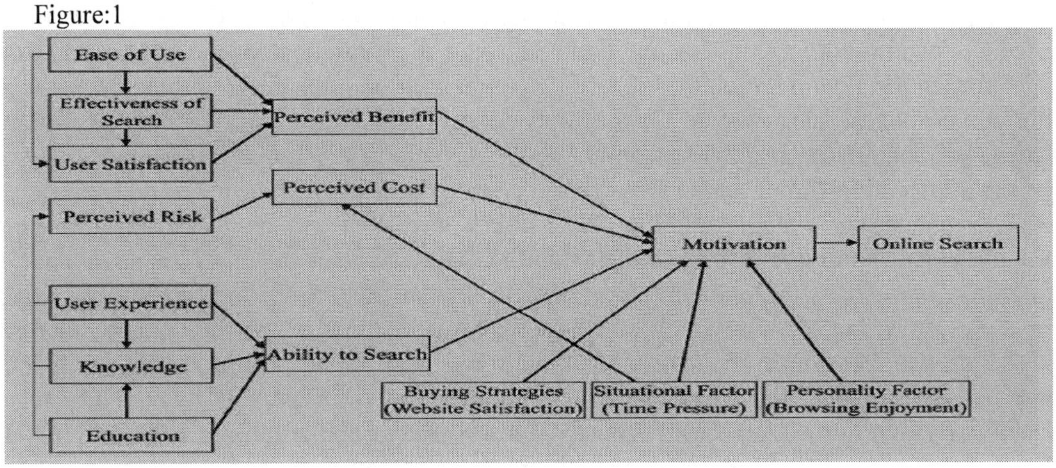

Most of the millennials are at their 20s and 30s and have career and family who lives in their own homes. Most of this age group is employee who is task-oriented and they prefer to avoid unnecessary conversations and want to make things done in an easy and quicker way (Howe, 2015). 28 percent of Indian millennials make purchase due to the social media recommendations and about 63 percent of Indian millennials stay updated on their favorite brands through social media (Deloitte India Survey, 2019).

Studies also show that driving forces in the consumers market have also changed with the changing generations. Price of products, Taste and Convenience are the traditional drivers in the market. But today consumers are driven more by the emerging value drivers like Health & Wellness, Safety, Social Impact and Experience. Today the consumers want to know more about the source of the raw materials and about the farms and factories from where the product has been sourced. They want to know the complete story behind the journey of a product and company's practices while manufacturing a product. And thus a brand image has been created on the consumer's mind. This brand image has made its consumers loyal and this loyalty leads to repeat visits and purchases of the products. These consumers not only act as a brand advertisers but also promote the brand through word of mouth.

Female millennials occupy a larger part of this new generation consumers. The female millennials are considered as a new era of talent. millennials have a different life style and in case of the female millennials, their purchasing behavior has changed after joining the workforce. Roy & Saha(2007) stated in their study that the perception of female millennials towards fashion clothing as well as clothing industry has changed. It will transform the market of women wear from unorganized sector to organized sector.

Compared to those older generation consumers, shipping and low prices are not much important for the millennials. Their priority is convenience and flexible return in case of shopping. Same day delivery is also an important factor for the millennials and they expect such immediacy from the onlne channel as they get this from buying in stores (Forbes, 2019).

Millennials are not loyal towards any product or services (Gurau,2012, Myers and Sadaghiani,2010). They are not limited to one digital communication service. Millennials have numerous digital service options and go with the service where it's easy to work with. Millennials always look for the easiest way to get things done. But at the same time it does not mean that the millennials are consuming less as they are engaged with more than one digital service. This generation consumers are more flexible than that of the other consumers group.

4. Conclusion

Since the millennials are comfortable with the new technology advancements and are more engaged in social media and hence are updated about the activities around the world. This new generation people are more into job or engaged in their own business so they always seek for the quickest and easiest way to make things done. It has been observed from the studies that there is a significant increase of online shopping among the Indian millennials. Online shopping or e-shopping has become a new trend among the millennials where they can access every informations by a single click on their smart phones or any devices from anywhere around the globe at any time. Besides, the available online services for 24×7 hrs have made it much easier for the busy millennials to buy at their convenience. Since the Indian millennials are in their younger age group which constitutes a larger population of the country, there is larger scope for the e-retailers to establish their business in the country. With a higher disposable income compared to the previous generations the millennials buying behaviour will have a positive impact on the economy. And for this, a regular study on the millennials buying behaviour is very important. It was found that keeping in mind both the price factor and convenience level, today's Indian millennials decide their buying decisions. There is a scope for the Millennials to prefer the integration of both online and offline stores to experience a seamless shopping in India.

References

[1] Aleksić, D., & Rangus, K. (2020). Millennials. *Encyclopedia of Creativity*, 157–163. https://doi.org/10.1016/b978-0-12-809324-5.23869-8

[2] Bansal, N., & Srivastava, D. (2017). Motivation & Attitude of Generation Y in India: An Exploratory Study. *Indian Journal of Industrial Relations*, *53*(1), 102.

[3] Bigdeli, O., Afaghzadeh, S., Albadvi, A., & Abdolvand, N. (2009). Barriers of online shopping in developing countries: Case study Iran. *Proceedings of the IADIS International Conference*

E-Commerce 2009, Part of the IADIS Multi Conference on Computer Science and Information Systems, MCCSIS 2009, (April), 3–10.

[4] Bucko, J., Kakalejčík, L., & Ferencová, M. (2018). Online shopping: Factors that affect consumer purchasing behaviour. *Cogent Business and Management*, *5*(1), 1–15. https://doi.org/10.1080/23311975.2018.1535751

[5] Burke, R. J., Cooper, C. L., & Antoniou, A. S. G. (2015). The Multi-generational and Aging Workforce: Challenges and Opportunities. In *The Multi-generational and Aging Workforce: Challenges and Opportunities*. https://doi.org/10.4337/9781783476589

[6] Cimperman, M. B., Harrison, A., Hatch, J., Pillar, K., & Snipes, T. (2018). Understanding the Millennial Buying Process in the Modern Digital Era -Secondary Research. *Honors Research Projects*, 546.

[7] Childs, R. D., Robinson, P., Mcgovern, T. M., & Gingrich, G. (2015). The millennial generation. *Transforming American Governance: Rebooting the Public Square*, 307–321. https://doi.org/10.1177/2158244017697158

[8] Devi, M., Das, L., & Baruah, M. (2018). Online shopping behaviour of the consumers in Jorhat town of Assam. *Asian Journal of Home Science*, *13*(2), 538–544. https://doi.org/10.15740/has/ajhs/13.2/538-544

[9] Deloitte. (2019). The Deloitte Global Millennial Survey 2019: Societal discord and technological transformation create a "generation disrupted." *Deloitte*, 31. Retrieved from https://www2.deloitte.com/content/dam/Deloitte/global/Documents/About-Deloitte/deloitte-2019-millennial-survey.pdf

[10] Deloitte. (2019). Unravelling the Indian Consumer, (February), 1–31. Retrieved from https://www2.deloitte.com/content/dam/Deloitte/in/Documents/consumer-business/Unravelling the Indian Consumer_web.pdf

[11] Jain, R. (2018). Techno-Savvy Millennials : A Booming Consumer Segment of India. *Effulgence-A Management Journal*, *16*(Spl2), 32. https://doi.org/10.33601/effulgence.rdias/v16/ispl2/2018/32-39

[12] Jain, R. (2015). Growth Drivers of Online Shopping in Small Cities of India, *7782*, 80–87

[13] Jun, G., & Jaafar, N. I. (2011). A Study on Consumers' Attitude towards Online Shopping in China. *International Journal of Business and Social Science*, *2*(22), 122–132.

[14] Katwani, A. (2011). A Comparative Study of Online Shopping and Traditional Shopping. *Journal on Contemporary Issues of Law*, *3*(8), 1–10.

[15] Kiran, B. S. (2019). Comparative Study on Online and Offline Shopping. *International Journal for Research in Applied Science and Engineering Technology*, *7*(10), 174–178. https://doi.org/10.22214/ijraset.2019.10029

[16] Klapilova Krbova, P. (2016). Generation Y Attitudes towards Shopping: A Comparison of the Czech Republic and Slovakia. *Journal of Competitiveness*, *8*(1), 38–54. https://doi.org/10.7441/joc.2016.01.03

[17] Kulviwat, S., Guo, C., & Engchanil, N. (2004). Determinants of online information search: A critical review and assessment. *Internet Research*, *14*(3), 245–253. https://doi.org/10.1108/10662240410542670

[18] Kumari, S., & Mishra, N. (2015). Countries A Comparative Study, (7), 1–4.

[19] Licsandru, T. C., & Cui, C. C. (2019). Ethnic marketing to the global millennial consumers: Challenges and opportunities. *Journal of Business Research*, *103*(February), 261–274. https://doi.org/10.1016/j.jbusres.2019.01.052

[20] Liu, J., Zhu, Y., Serapio, M., & Cavusgil, S. T. (2019). The new generation of millennial entrepreneurs: A review and call for research. *International Business Review*, *28*(5), 101581. https://doi.org/10.1016/j.ibusrev.2019.05.001

[21] Moreno, F. M., Lafuente, J. G., Carreón, F. Á., & Moreno, S. M. (2017). The Characterization of the Millennials and Their Buying Behavior. *International Journal of Marketing Studies*, *9*(5), 135. https://doi.org/10.5539/ijms.v9n5p135

[22] Muda, M., Mohd, R., & Hassan, S. (2016). Online Purchase Behavior of Generation Y in Malaysia. *Procedia Economics and Finance*, *37*(16), 292–298. https://doi.org/10.1016/s2212-5671(16)30127-7
[23] Naim, M. F., & Lenka, U. (2018). Development and retention of Generation Y employees: a conceptual framework. *Employee Relations*, *40*(2), 433–455. https://doi.org/10.1108/ER-09-2016-0172
[24] Ordun, G. (2015). Millennial (Gen Y) Consumer Behavior Their Shopping Preferences and Perceptual Maps Associated With Brand Loyalty. *Canadian Social Science*, *11*(4), 1–16. https://doi.org/10.3968/pdf_294
[25] Pandey, S., & Chawla, D. (2014). E-lifestyles of Indian online shoppers: A scale validation. *Journal of Retailing and Consumer Services*, *21*(6), 1068–1074. https://doi.org/10.1016/j.jretconser.2014.06.012
[26] Parment, A. (2013). Generation Y vs. Baby Boomers: Shopping behavior, buyer involvement and implications for retailing. *Journal of Retailing and Consumer Services*, *20*(2), 189–199. https://doi.org/10.1016/j.jretconser.2012.12.001
[27] Profile, S. E. E. (2014). IN MANAGEMENT & SOCIAL SCIENCE Empyreal Institute of Higher Education, *2*(August).
[28] Purani, K., Kumar, D. S., & Sahadev, S. (2019). e-Loyalty among millennials: Personal characteristics and social influences. *Journal of Retailing and Consumer Services*, *48*(July 2018), 215–223. https://doi.org/10.1016/j.jretconser.2019.02.006
[29] Raina, R. (2017). Engaging Indian Millennial @ Workplace.
[30] Rajput, N., & Khanna, A. (2014). Dynamics of Young Indian Consumers' Buying Behaviour Towards Branded Apparels: Gender Perspective. *Archives of Business Research*, *2*(5), 84–106. https://doi.org/10.14738/abr.25.596
[31] Sarkar, R., & Das, D. S. (2017). Online Shopping vs Offline Shopping : A Comparative Study. *2017*, *3*(1), 1367–1368. Retrieved from http://ijsrst.com/paper/841.pdf
[32] Salim, M., Alfansi, L., Darta, E., Anggarawati, S., & Amin, A. (2019). Indonesian Millenials Online Shopping Behavior. *International Review of Management and Marketing*, *9*(3), 41–48. https://doi.org/10.32479/irmm.7684
[33] Sethi, R. S., Kaur, J., & Wadera, D. (2018). Purchase Intention Survey of Millennials. *Academy of Marketing Studies Journal*, *22*(1), 16.
[34] Sharma, V. (2015). Experience-Role of Youth and, (04), 323–330.
[35] SivaKumar, A. K., & Gunasekaran, A. (2017). An Empirical Study on the Factors Affecting Online Shopping Behavior of Millennial Consumers. *Journal of Internet Commerce*, *16*(3), 219–230. https://doi.org/10.1080/15332861.2017.1317150
[36] Sivanesan, I. R., & Monisha, I. C. (2017). Comparative Study on Factors Influencing Online and Offline Shopping (With Special Reference to Kanyakumari District of Tamil Nadu). *International Journal of Research in Management & Business Studies*, *4*(1). Retrieved from www.ijrmbs.com
[37] Silpa, K. S., Rajasree, P. U., & Balasubramanian, D. P. (2016). A Study on Peoples' Perceptions towards Online Shopping. *Bonfring International Journal of Industrial Engineering and Management Science*, *6*(3), 93–95. https://doi.org/10.9756/bijiems.8187
[38] Sun, T. R. (2015). The Review of Physical Store Factors That Influence Impulsive Buying Behavior, *2*(9), 1048–1054
[39] S. Nethra; Dr. V. T. Dhanaraj. (2016). Consumer Behaviour and Attitude Towards Online Shopping. *International Journal of Applied and Advanced Scientific Research*, *1*(I), 170–173.
[40] Sureshkumar, R. (2017). Rural Consumer Attitude Towards Online Shopping : an Empirical Study of Rural Area. *International Journal of Innovative Research in Management Studies*, *1*(12), 1–5.
[41] Taufique, K. M. R., & Vaithianathan, S. (2018). A fresh look at understanding Green consumer

behavior among young urban Indian consumers through the lens of Theory of Planned Behavior. *Journal of Cleaner Production*, *183*, 46–55. https://doi.org/10.1016/j.jclepro.2018.02.097

[42] The Rural. (2017). *The Rural*. https://doi.org/10.4324/9781315237213

[43] Yadav, R., & Khandai, S. (2019). Buying Behaviour of Millennial Women : A Cross-Regional Analysis An Literature Survey of Studies Done About Women Managers on Select Countries with Special ... Buying Behaviour of Millennial Women : A Cross-Regional Analysis. *International Journal of Applied Business and Economic Research*, (November).

[44] Yongo, E., Manyala, J. ., Kito, K., Matsushita, Y., Outa, N. ., & J.M, N. (2016). Diet of Silver Cyprinid, Rastrineobola argentea in Lake Victoria, Kenya. *International Journal of Advanced Research*, *4*(4), 144–149. https://doi.org/10.21474/IJAR01

Mitigating Tourism Impact through Ecological Ethics: Conceptual Approach

Gebi Basar

Research Scholar, Department of Humanities & Social Sciences, NERIST, Nirjuli, 791109, Arunachal Pradesh, India. E mail- gebibasar@gmail.com

Mudang Tagiya

Research Scholar, Department of Humanities & Social Sciences, NERIST, Nirjuli, 791109, Arunachal Pradesh, India. E mail- tagiyam@gmail.com

P.Parida

Professor, Department of Humanities & Social Sciences, NERIST, Nirjuli, 791109, Arunachal Pradesh, India. E mail- pra.parida@gmail.com

Abstract. Today among the business industries in the world, tourism is the fastest growing business industry. Tourism is considered as one of the most profitable business among the local, national, and international business industries. It is considered as region's economic catalyst. But tourism may be disastrous for the natural environment and ecology when it is unsustainable. The region which crosses the legal and ethical barriers in tourism industry to earn more profit, can lead to an enormous degradation of the environment, in that particular area. Due to such irresponsible and unsustainable tourism, the local population of human, flora, and fauna, suffer greatly. It gives a great stress on use of community and regional land and can lead to encroachment of land, erosion of soil, over pollution, loss of natural ecology, and might result in the extinction of the endangered species. Unsustainable tourism might lead to the demolition of its resources, ecology of many species and the natural environment. Ecological ethics is an emerging concept in the philosophical field of studies of the moral values, conciseness and relationship of human beings with respect to its non- human contents in the nature. In the discipline of environmental studies, ecological ethics is one of the very important component which sets up the connection between the human being and mother earth. This is an exploration to study, which is conceptual in nature with logical debate. The paper proposes the significance of ecological ethics and its relevance in the field of tourism. The study discusses various adverse impacts of tourism on ecology and natural environment. It also tries to understand how the implication of ecological ethics could mitigate these impacts and maintain the ecological balance.

1. Introduction

Tourism promotes good will among the people of different places, countries or regions. It encourages international understanding. It creates numerous social and economic benefits for the countries, who are involved in tourism.

Among the world's largest industries, tourism and travel industry is extremely huge. Tourism provides employment to the locals of the certain tourist destinations as it involves the activities such as, promotion of local handicrafts, traditional & cultural activities, local services etc. resulting in the generation of income as well as the foreign exchange. Hence, it creates prosperity and goodwill among the people of the nations.

However, people tend to forget some basic ethics which might lead to some negative effects on our nature. Despite of its numerous benefits, excess tourism may lead to imbalanced in the ecology of many species including the human being.

According to the United Nation's world tourism organization, there is excessive exports of oils, food products and automobiles in past few years. The entering tourism boosts the economic growth of a region's economy while providing high prospective to support job creation. The outward departing tourism promotes cross-cultural understanding and goodwill. According to Mark Twain, "Travel is fatal to prejudice, bigotry, and narrow- mindedness..." (USA Today.com)

The unsustainable tourism might create a huge pressure on local and regional land uses, which might increase the erosion of soil, increase pollution, loss of natural ecology, and more pressure on the species which are already declared endangered. These can slowly result in the adverse negative consequences and damage in the environment and its resources on which tourism itself depends (Worldcounts.com).

Too many and increasing tourists might lead to an adverse impact on the environment and might degrade the standard of the natural existence. This occurrence might create load & pressure in the natural ecology of certain destinations such as in rural areas of Patagonia, Chile and urban areas of Barcelona and Spain, which are known as places burdened by excess tourism.

Another such example is the mountain slopes and valleys of Everest which are littered with rubbish and wastes. In the tourist destinations such as Iceland, the population of tourists has outnumbered the population of local residents.

The top 10 parks in U.S.A reported more than 44 million visitors in the year 2016. The National Park Service is trying to look for solutions to protect natural resources as the visitors are increasing year- on-year basis.

Worldwide tourism accounted 8% of global greenhouse gas emissions in the year 2009 to 2013. New researches done in past few years has found that the tourism sector acts as a bigger polluter than the construction industry (www.carbonbrief.org).

The philosophical field that considers the moral relations, values and consciousness between human beings and their natural ecology and environment is known as ecological ethics. It regards that, the human beings must consider certain commitments and sense of responsibility towards its nature. It looks about to assist the individuals and their leaders to become aware of ecologically conscious and to act responsibly when they do things that affect the environment.

Considering human being as the most important entity over other beings and natural environment is one of the major reason for the exploitation of natural resources and minerals in the name of mankind development. It takes millions of years for nature to recover the natural resources and minerals but it takes only few hours to use and burn it. In fact, human beings are gaining less and losing more.

Today many people acknowledge that some collaborative instructions or general rules should exist between people when they interact with nature. If they don't do so, nothing in our lives would be foreseeable and safe. Besides actual laws, it is important to know that, there are some basic common ethics or principles of what is right and what is wrong that everyone should agree upon and must follow and live by it. As a specialized part of ethics, ecological ethics is concerned with the human morality of treating the nature in a right way. As, treating it in a wrong way may affect the ecology and the natural environment. The growing human needs have troubled these natural components and created an ecological imbalance. Expanding and growing population, emission of harmful gases and the thought & philosophy of anthropocentrism among the people where, human beings are considered as the most superior being over nature has posed a serious threat to the ecological balance. Indirectly or directly, unsustainable tourism without any ethical norms is one of reason for all these increasing environmental problems which might destroy the existence of not only the human beings but the whole nature. Therefore, it is very important and necessary for us to realize the importance of sustainable alternatives and ecological ethics not only in tourism industry but in other industries too.

Increasing tourism increases the population in the particular tourist destinations resulting in the increase of cutting down of forests for creating hotels & tourist lodges, parks etc., increases the movement of automobiles, increases air and water pollution, increases the use of plastics and wastes etc. which disturbs the natural habitat of many species of animals and plants as well. And which may affect human beings in a long run.

Following are some data about the domestic and international arrivals of tourists from different places to India and its north-eastern states which clearly shows how the population of human beings are increasing each year in these places.

Table no.1 shows the number of foreign tourist arrivals in N.E states of India and in all India.

Table No.1: Foreign Tourist Arrivals

Year	Arunachal Pradesh	Assam	Manipur	Meghalaya	Mizoram	Nagaland	Sikkim	Tripura	N-E Region	All India
2005	289	10782	316	5099	273	883	2677	16523	36842	9939782
2006	607	10374	295	4287	436	1002	3245	18026	38272	11403661
2007	2212	12899	396	5267	669	936	17498	3181	43058	13267273
2008	3020	14426	354	4919	902	1209	19154	3577	47561	14112590
2009	3945	14942	337	4522	513	1423	17730	4246	47658	14372300
2010	3395	15157	389	4177	731	1132	20757	5212	50950	17910178
2011	4753	16400	578	4803	658	2080	23602	6046	58920	19497126
2012	5135	17543	749	5313	744	2489	26489	7840	66302	18263074
2013	10846	17638	1908	6773	800	3304	31698	11853	84820	19951026
2014	5204	21537	2769	8664	836	3679	49175	26688	118552	22567650
2015	5705	24720	3260	8027	798	2769	38479	34886	118644	23326163
2016	6598	12685	3064	8476	942	3260	66012	36780	137817	24707732

Source: NEDFi Databank NEC, Govt. of India

Table no. 2 showcases the number of domestic tourist arrivals in the N.E states of India with comparison with all India Statistics.

Table No. 2. Domestic Tourists Arrival in NE States in comparison with All India statistics

Year	Arunachal Pradesh	Assam	Manipur	Meghalaya	Mizoram	Nagaland	Sikkim	Tripura	N-E Region	All India
2005	50560	2467652	94299	375901	44715	17470	251744	216330	3520676	391948589
2006	80137	3268657	116984	401529	50987	15850	292486	230645	4459281	462310177
2007	91100	3436833	101484	457685	43161	22085	329075	244795	4728225	526564364
2008	149292	3617306	112151	549936	55924	21129	460564	245438	5213748	562982298
2009	195147	3850521	124229	591398	56651	20953	615628	317541	5774077	668800482
2010	227857	4050924	114062	652756	57292	21094	700011	342273	6168279	747703380
2011	233227	4339485	134505	667504	62174	25391	552453	359515	6376265	864532718
2012	132243	4511407	134541	680254	64249	35915	558538	361786	6480945	1045047536
2013	125461	4684527	140673	691269	63377	35638	576749	359586	6679293	1145280443
2014	335974	4826702	115499	716469	68203	58507	562418	361247	7045019	1281952255
2015	352067	5491845	146169	751165	66605	64616	705023	363172	7940662	1431973794
2016	385875	5160599	150638	830887	67238	58178	747343	370618	7771376	1613551505

Source: NEDFI Databank NEC, Govt. of India

2. Objective of the study

i) To study the various adverse impacts of tourism with special reference to the environment impact.

ii) To study ecological ethics as a tool to mitigate the adverse impact of tourism on environment.

3. Review of Literature

The damage created by human being to its nature is irreparable according to (Barta, 2011). The concern and the relation between human and nature has emerged since very early days when the great philosopher like John Muir (1838-1914) and Jeremy Bentham (1748-1832) stepped in field of philosophy. And till today they both are still very influential in the philosophical field of environmental ethics. Similarly, Aldo Leopold's work during (1949) 'A sand country almanac' cannot be ignored where he did a significant work related to ethics related to land.

Rachel Carson's "Silent Spring" published in 1962, was another contribution in the field of environmental studies and was a development of such awareness among the people and its readers, how the widespread use of chemical pesticides was creating a serious threat to the health of common public and leading to the demolition of nature and wildlife.

The papers published by Lynn White's "The Historical Roots of our Ecologic Crisis" (March 1967) in field of science had a very crucial significant where, he believed man's relationship with the natural environment was always a dynamic and interactive one, even in the middle ages, but marked the industrial revolution was a fundamental turning point in our ecological history. Also, the write up by Garrett Hardin's book "The Tragedy of the Commons" (1968), which was an eye opener to "the damage that innocent actions by individuals can inflict on the environment". He also introduced, Hardin's first law of human ecology.

One of the significance report was Paul Ehrlich's 1968 book, "The Population Bomb" which alerted the disastrous consequence of the twisting human population has on the earth and natural resources. During 1970s, in field of science, the study of ethical questions raised by human relations with the non-human environment emerged as an important subfield of philosophy. An essay "Exploring New Ethics for Survival" by Garett Hardin's (1972) was also an important and influential work in the field of research.

Naess (1973) founded the theory "Deep Ecology" to refer to an environmentalism that believes fundamental changes in the way our species conceives our relation to nature are necessary before we find a way out of the ecological crisis around us. According to him it is a normative political philosophy which contrasted with shallow ecology. Which also assumes that minor technical fixes in our present way of doing things will solve our environmental problems. He believed, with our present way of considering the nature only as a resource will never work according to deep ecology.

Holmes Roston III (1975) introduced the concept of "ecological consciousness" in his paper "Is there an ecological ethic?" Nature must be appreciated for its intrinsic value, or worth in itself, regardless of how humans might benefit from it. Further researches and academic journals by many researchers were reviewed, and the term "Environmental Ethics" 1978 was founded by Eugene Hangrove.

From 1980's onward, research, publications and teachings in the philosophical field of Environmental Ethics was expanded rapidly. And at present it is a flourishing area of research. Extending the further research and theory of Deep Ecology of Arne Naess, Devall and Sessions 1985, 75 quoted, "With maturity, human beings will experience happiness when other life forms experience happiness and sorrow when other living beings and forms experiences sorrow. We feel sad when our brother or a dog or a cat feels sad, but we will grieve when living beings, including landscapes, are destroyed. In our civilization, we have vast means of destruction at our disposal but extremely little maturity in our feeling. Only a very narrow range of feelings have interested most human beings until now" Hangrove (1992) published a book "Foundation of Environmental Ethics" He starts with the history of Greek philosophy related to nature and environment. He termed applied ethics along with environmental ethics in his write-ups. He proposed the study of philosophical aspects and social attitudes in western culture with respect to its environment and nature, including the protection and conservation of wildlife, land management and uses etc. He gave the ontological argument for preservation of nature. He examined the historical significance and prepared a framework for further discussions on environmental ethics.

In the book by Callicott (1994) "Earth's Insight: A multicultural survey of ecological ethics from the Mediterranean Basin to Australian outback" examined that the environmental disaster is a global problem, yet ecological consciousness is merely discussed in the felid of philosophy and it is mostly centered in western philosophy and religion. He further expanded the scope of environmental ethics, implanted in non-western worldviews, to include the teachings on ecological consciousness. He explored broadly the sacred texts of Islam, Zen Buddhism, Jainism, Taoism, Hinduism, Confucianism, and, as well as the oral traditions of North and South America, Australia and Polynesia. Also, he documented the various attempts made by peoples to practice the environmental ethics in their locality and surroundings. He finally came up with a question to all human sharing the fate of our small planet which is of a very crucial significance. He also questioned, how the world's diverse environmental studies and philosophies can be brought together in a complementary and congruous fullness.

Rosa and Silva (2005), in their Journal of Agricultural and Environmental Ethics focused on the conservation of various species and ecology in the European Union. They analyzed the concept and values in their questions regarding the term like environmental ethics, anthropocentrism, non-anthropocentrism, biocentrism and egocentrism. They examined the value and concept considering the following categories such as,

- Human worth and morals.
- asset worth and values e.g. economy, income, market, costs and benefits
- private property e.g. non-interference, property rights, landowner rights
- natural resources e.g. production, consumption, management of resources
- cultural values e.g. aesthetic value, scientific value, heritage
- sustainability e.g. future generations, sustainable development, sustainable use
- stewardship e.g. nature preservation, restoration, care, management);
- living beings e.g. animal rights/welfare, biological interests, individual organisms
- ecological values e.g. ecosystem stability/equilibrium, ecological services
- Intrinsic value of nature etc.

Fairbanks (2010) "Environmental Goodness and the Challenge of American Culture" where he wrote in her journal the western morality has never recognized nature-focused morality. As in the past western regions and philosophers always encouraged the idea of that human superiority over the natural environment and there is no ethical code to control our relationship to Mother Nature. According to her there must be a complete change in our viewpoint towards the nature if we have to reduce and match up the challenges created by present environmental issue

4. Research Methodology

The study aims to understand ecological ethics as a tool to mitigate adverse impact of tourism on the natural ecology. This study is a conceptual approach base on secondary information collected from various sources. The paper discusses various adverse impacts of tourism on natural ecology and need of ecological ethics in the field of tourism.

5. Analysis and observation

5.1. Analysis-I

The following table no.3 throws the light on the various impacts of tourism activities on environment.

Table no. 3 - Tentative lists of negative impacts of tourism on environment

Sl. No.	Negative outcome of tourism activities	Source
1	Deforestation	Kuvan, Y. (2010)
2	Ecological imbalance	Parsons, E. C. M. (2012)
3	Wildlife disturbances	Green, R., & Giese, M. (2004)
4	Effect on Land	Boori, M. S., Voženílek, V., &Burian, J. (2014)
5	Environmental Impacts of Tourism on Lakes.	Dokulil, M. T. (2013)
6	Global Carbon foot prints	Lenzen, M. et al. (2018)

7	Negative Impact of Tourism on Mountains	Cioancă, L. M. (2015, June)
8	Air Pollution	Andereck (1993), Lee et al. (2009).
9	Water Pollution	Lee et al. (2009)
10	Solid wastes	(UNEP, 1997)
11	Biodiversity loss	Joseph (2015)

As we know that, 'every coin has its flip side', indeed there is no doubt that tourism has got several benefits and contributing to the economy of the region and nation as a whole. However, one cannot deny the negative outcome of the tourism, which needs to be understood and work on it. From the table 3, it is clearly observed that various areas are affected and various environmental problems are contributed by the tourism. One cannot compromise with the destruction of the environment at the exchange of benefits. The above matters are serious and need to be pondered and discussed upon.

5.2. Analysis-II

Only in the past few decades of the twentieth century, the ecological consciousness is considered as a recent development in the field of philosophy and is also considered as a sub topic of organized ethical knowledge. Global climate change, deforestation, over population, degradation of natural resource, extinction of endangered species are the common issues among the adverse impact of tourism. Ecological ethics and consciousness are main attributes of environmental studies which set ups the relationship between humans and the natural environment. With ecological ethics in tourism industry, one can ensure that he/she is doing his/her duty to keep the nature protected and safe. Any moment a forest is cleared, which are becoming scanter to find, to construct hotels, recreation centers, amusement parks and other means of human entertainment. With the speedy increase in travel and tourism industry, the human population increases in the particular tourist destinations and hence, utilization of natural resources also increases. It has caused the depletion of our planet's ability to provide the natural resources. The exhaustion rate of natural resources is growing faster than the rate of natural replenishment. There can be an improved human values, morals and principles if, ecological ethics and consciousness is constructed on the basis of scientific understanding, which may also help to make wise decisions to conserve the entire human race, the ecology and the earth.

6. Conclusion

Ecological ethics plays an important role in defining the human-nature relationship. These ethics help in finding a formula of how humans behave towards the natural environment. Applying these formulas in long-term sustainable development plans, can bring significant economic benefits for investments, and more importantly, protect natural resource for future generations.

This is a bitter reality that the benefits of tourism industry can be a gain for few, but its negative impacts affect everyone and the natural environment. And thus, the ecology of the many species, which is not off course ethically correct. Therefore, in order to solve these problems, certain mechanisms are needed to be taken by the society and authorities to controland mitigate these adverse effects. In this research, after introducing the ethical aspects of natural environment in tourism, some measures for reducing and controlling the negative effects of tourism were reviewed and introduced, so that the industry can move towards social and environment equity and ultimately, be ethical and beneficial for all.

References

[1] Andereck, K. L. (1993). The impacts of tourism on natural resources. Parks & Recreation (Arlington), 28(6), 26-86.

[2] Barta, A. (2011), Caiet de lucrări practice, Presa Universitară Clujeană, Cluj-Napoca.

[3] Bentham, J. (1996). The collected works of Jeremy Bentham: An introduction to the principles of morals and legislation. Clarendon Press.

[4] Bentham, J., & Bowring, J. (1843). The Works of Jeremy Bentham (Vol. 7). W. Tait.

[5] Boori, M. S., Voženílek, V., & Choudhary, K. (2014). Land use/cover disturbance due to tourism in Jeseníky Mountain, Czech Republic: A remote sensing and GIS based approach. The Egyptian journal of remote sensing and space science, 18(1), 17–26.

[6] Callicott, J. B. (1984). Non-anthropocentric value theory and environmental ethics. American Philosophical Quarterly, 21(4), 299-309.

[7] Callicott, J. B. (1990). Whither conservation ethics? Conservation Biology, 4(1), 15-20.

[8] Callicott, J. B. (1994). Earth's insights: a multicultural survey of ecological ethics from the Mediterranean Basin to the Australian Outback. University of California Press.

[9] Carson, R. (2009). Silent spring. 1962.

[10] Cioancă, L. M. (2015, June). The Negative Impact of Tourism in The Area of The Bârgău Mountains onto The Local Environment. In International Conference Knowledge-Based Organisation (Vol. 21, No. 1, pp. 170-178). De Gruyter Open.

[11] Devall, B. (1980). The deep ecology movement. Nat. Resources J., 20, 299.

[12] Devall, B. (1982). John Muir as deep ecologist. Environmental Review: ER, 6(1), 63-86

[13] Dokulil, M. T. (2013). Environmental Impacts of Tourism on Lakes. In Eutrophication: causes, consequences and control 81-88.

[14] Ehrlich, P. R. (1978). The population bomb.

[15] Ehrlich, P. R., & Ehrlich, A. H. (2009). The population bomb revisited. The electronic journal of sustainable development, 1(3), 63-71.

[16] Fairbanks, S. J. (2010). Environmental goodness and the challenges of American culture. Ethics & Environment, 15 (2), 79-102.

[17] Green, R., & Giese, M. (2004). Negative effects of wildlife tourism on wildlife. Wildlife Tourism: Impacts Management and Planning. CRC for Sustainable Tourism and Common Ground Publishing, Altona, Vic, 81-97.

[18] Hardin, G. (1968). The tragedy of the commons. science, 162(3859), 1243-1248

[19] Holmes Rolston III Ethics: An International Journal of Social, Political and Legal Philosophy 18(no. 2, 1975):93-109. Is There an Ecological Ethic?

[20] Hargrove, E. C. (1992). Foundations of environmental ethics.

[21] Joseph FN (2015). Effect of tourism on biodiversity in the coastal environment; A study of Bentota area in the south–western coastline in Sri Lanka. In: Proceedings of the International Postgraduate Research Conference 2015 University of Kelaniya, Kelaniya, Sri Lanka, (Abstract), 339 pp.

[22] Kuvan, Y. (2010). Mass tourism development and deforestation in Turkey. Anatolia, 21(1), 155-168.

[23] Lee H, Baylon Verances J, Song W (2009) The tourism-environment causality. International Journal of Tourism Sciences 9(3):39–48

[24] Lenzen, M., Sun, Y.-Y., Faturay, F., Ting, Y.-P., Geschke, A., & Malik, A. (2018). The carbon footprint of global tourism. Nature Climate Change, 8(6), 522–528.

[25] Leopold, A. (1933). The conservation ethic. Journal of Forestry, 31(6), 634-643. Leopold, A. (1968). From A Sand County Almanac (1949). Oxford.

[26] Mieczkowski, Z. (1995). Environmental issues of tourism and recreation. University Press of America.

[27] Muir, J. (1979). John of the mountains: The unpublished journals of John Muir. University of Wisconsin Press.

[28] Parsons, E. C. M. (2012). The negative impacts of whale-watching. Journal of Marine Biology, 2012.

[29] Rosa, H. D., & Da Silva, J. M. (2005). From environmental ethics to nature conservation policy: Natura 2000 and the burden of proof. Journal of Agricultural and Environmental Ethics, 18(2), 107-130.

[30] Sadeghian, M. M. (2019). Negative Environmental Impacts of Tourism, A brief review.

[31] Sunlu, U. (2003). Environmental impacts of tourism. In Conference on the Relationships between Global Trades and Local Resources in the Mediterranean Region (pp. 263-270).

[32] UNEP. 1997. Environmental Good Practice in Hotels, Case Studies (jointly with IHRA).

[33] White, L. (1967). The historical roots of our ecologic crisis. Science, 155(3767), 1203-1207.

[34] Whitney, E. (1993). Lynn White, Eco theology, and history. Environmental Ethics, 15(2), 151-169.

[35] https://www.carbonbrief.org/tourism-responsible-for-8-of-global-greenhouse-gas-emissions-study-finds

[36] https://www.conservation.org/blog/plastic-pollution-and-over-tourism-its-not-totally-your- fault

[37] https://www.unenvironment.org/news-and-stories/story/paradise-lost-travel-and-tourism-industry-takes-aim-plastic-pollution-more

[38] https://traveltips.usatoday.com/positive-negative-effects-tourism-63336.html

[39] http://www.scienceclarified.com/El-Ex/Environmental-Ethics.html

Impact of HR audit on Organisational Performance: A study on Power sector Organisations in Arunachal Pradesh, India

*Gautam Ku. Roy[1], Manmohan Mall[2] and Prataprudra Parida[3]

North Eastern Regional Institute of Science & Technology, Nirjuli, Arunachal Pradesh, India

Abstract:
The phenomenal growth in Indian power sector during the last two to three decades is accompanied by various challenges across the board. The initial growth in this sector is rooted to large scale investment, timely project execution and performance in the long run pose various kinds of challenges that need to be addressed sustainably. Even technology also plays a crucial role in this sector. Hence, the challenges ranging from attracting fresh pool of talents to regular upgradation of skill set among the existing personnel as well as enhancement of managerial, attitudinal and behavioural competencies are the major thrust areas where HR department is expected to lead efficiently in augmenting the performance of organisation. However, little is known in relation to how much the HR department/professionals are prepared to face these challenges. This research work tries to address the future issues through a detailed study on HR audit and its impact on organisational performance by utilising quantitative survey of managers as well as technical and non-technical employees (n = 120) of government undertakings in power sector. The findings indicate that, except compliance audit approach in HR audit other approaches i.e. productivity and value-for-money audit were insignificant while determining their impact on organisational performance. Hence, it is recommended to take the full advantage of the abilities of their employees by auditing human resource productivity in order to improve the performance of the organisation.

Keywords: Human Resource Audit, Organisational Performance, productivity audit, Compliance Audit, value-for-money audit.

1. Introduction:

The origin of modern management practices can be traced back to 16th century, when low level of efficiency was the major cause for failure of certain enterprises (Sir Thomas More, 1478-1535). According to Peter Drucker (2005) 'management' must be included as a factor in the production process along with materials and money. Management as a discipline is considered as an integrated function that encompasses corporate policy formulation and to achieve the stated policy objective along with organize, plan, control and direct the organization's resources efficiently. Among the various resources 'human resource' plays a prominent role towards achievement of organizational objectives and requires special attention in addition to other functions of management viz. marketing, finance, production etc. The people that work for an organization is referred to as human resource in management. For an organization to achieve its planned goal and for the human resource to fully benefit from working at the organization they need to be well managed. One such practice of human resource management is concerned with hiring and developing skilled employee who assist the business to achieve its target in an efficient and effective way. This has necessitated the organization to conduct human resource evaluation from time to time. Out of the several methods of human resource evaluation, the audit approach is concerned with evaluating the HR function effectiveness. In this context, 'auditing' is described as an investigative and information processing activity, which evolved in response to the need for independently, verified stewardship report (Parker, Ferris and Otley, 1989). The word audit comes from the Latin verb 'audire' (to listen), the auditors' role was to 'listen to the records' (Burrowes & Pearson, 2000). Burrowes and Pearsson give example of two scribes independently making records of what was consigned to a warehouse, with these being reconciled by a third and independent scribe, to ensure the integrity of the records. Auditing has evolved, and becoming increasingly specific, until the term functional audit has emerged. The HR audit is a functional audit. Thus, we can say that HR audit consists of diagnosing, analyzing, evaluating, and assessing future lines of action within the framework of human resource management. Authors like McConnell (2011), Carter, C. et.al. (2015) and Schuler, Budhwar and Florkowski (2002), Shaban, O. (2012) says that there is little difference between financial audit and auditing human resources except the involvement of people dimension in HR audit.

Energy sector plays a vital role in driving the socio-economic development of a country. The economic progress and demand for power are interrelated with each other. The economic progress leads to increased number of various economic activities, thereby leads to increased demand for electricity. In order to meet the additional demand capacity enhancement is necessary and thus a greater number of units need to be installed. The Indian power sector consists of electricity generation, transmission and distribution companies considered to be a most diversified sectors in the world. The power generation sources range from conventional like thermal, hydro and nuclear power to non-conventional like wind and solar. The rapid increase in electricity demand and more so that is expected in the years to come requires massive addition to the installed capacity. For instance, India has set an ambitious target of 100 GW of installed solar capacity within 2022 as shown in the below figure-1.

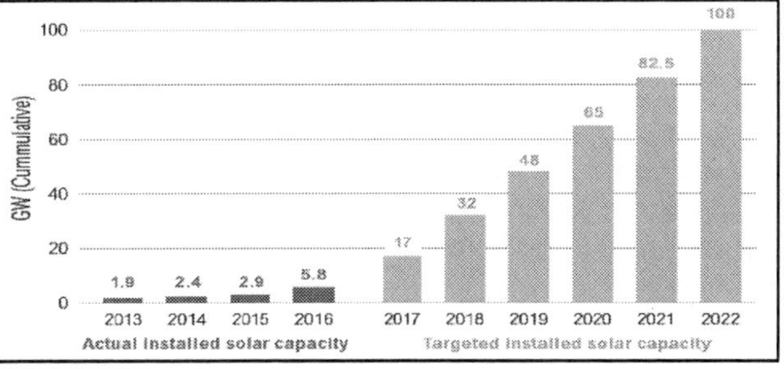

Figure-1: Targeted installed capacity in solar energy (Source: World Resources Institute)

The resultant of increase in installed capacity will require more manpower to operate. As per the report of working group on power in the 12th plan for an estimated capacity addition of 94,215 MW (including renewables) requires 407.67 thousand of manpower consisting of 312.92 thousand technical and 94.75 thousand non-technical. It was also estimated that the Manpower per MW ratio at the end of 12th Plan will be 4.74. Owing to the labour intensive feature of this industry, it is necessary for companies operating in this industry to manage their human resources effectively. Therefore, Human resource audit is one such tool that will assist the management in efficiently address and handle the issues concerning to manpower. The present research work was conducted on the assumption that various human resource audit approaches significantly impacts the performance of organisations operating in the power sector. The striking feature of power sector requires qualified manpower to operate the system efficiently. Hence, the importance of human resource department on this backdrop is evident because of the role it plays in respect of activities like-recruitment, selection, training, customer service, teamwork etc. The auditing of all these activities in a timely manner may lead to greater scale of achievements and hence better performance. Hence, the objective of this research work was to identify the impact of HR audit on organisational performance.

2. Literature Review:

There are different approaches of human resource audit which are used to provide different assurances and have different objectives. Human resource audit need not to be exhaustive, but should be focused on particular HR functions such as; training and development, performance appraisal, compensation and rewards and compliance. As per Devarajan Dasa, (2011), the primary goal of HR audit is to assess how efficiently the HR functions were performed by the HR department. (Olalla & Castillo, 2002) describe three approaches to HR audit such as legal approach, functional approach and strategic approach. While, Mock (2004); Rao T. V. (2014) described 04 types of HR audit such as compliance, best practice, strategic and function specific audits. In terms of Bhattacharyya (2007) HR audits can be performed on legal compliance, HR record keeping, employee relation audit and HR development audit. Nutley (2000), considered six 'Pure' audit types (Systems audit, compliance audit, Value-for-money audit; User satisfaction audit, value added audit and strategic contribution audit). Nutley correctly indicates that many HR audits are likely to be 'Hybrids' incorporating on the audit objectives and what is found during the fieldwork face. Similarly, (Yadav & Dabhade, 2014) discussed about 05 types of HR audit such as- functional audit, service audit, compliance audit, financial management audit and strategic audit. On the other hand, (Ukil, 2015; Kumari, 2017) suggested that auditors can adopt out of 05 approaches like-comparative approach, outside authority approach, statistical approach, compliance approach and MBO approach, any one for conducting HR audit. The discussions on various types of HR audit in the extant literature did not provide any agreement on the approaches. Hence, a summary of various audit approaches described in the literature and the related authors are summarized in Table-1. Where the audit approach is not specifically described by the author but can be implied from the commentary this element is also identified in the table-1.

Table-1: Author perspectives on various audit approaches

Sl No.	HR Audit approach	Author (Year and Country)
1	Compliance audit approach	Clardy (2004, Maryland); Nutly (2000, UK); Ollala & Castillo (2002, Spain); Rajesh K. Yadav, Nishant Dabhade (2014, India); Ukil, M. I. (2015, Bangladesh); Bhattacharyya (2007, India), Kumari (2017, India); Bloom, et al. (2006);
2	Value for money audit approach	Nutly (2000, UK); Ollala & Castillo (2002, Spain);
3	Productivity audit approach	Nutly (2000, UK); Ollala & Castillo (2002, Spain);
4	Financial audit	Bloom, et al. (2006);
5	Performance audit	Nutly (2000, UK); Ukil (2015, Bangladesh); Kumari (2017, India)
6	System audit	Nutly (2000, UK); Bloom, et al. (2006);
7	Functional audit	Ollala & Castillo (2002, Spain); Bloom, et al. (2006)
8	Value added audit	Nutly (2000, UK)

The primary objective of HR audit is to identify the adequacy, legality and effectiveness of HR department's detailed practice areas. The outcome of such review will assist in proper identification of HR practice gaps and later on a priority basis actions will be undertaken in order to minimize regulatory violations and lawsuits. It also improves job satisfaction and impacts positively on the performance of organisation. In line with the above discussions, this study has utilized 03 approaches to measure HR audit variable and tries to identify its impact on organisational performance.

2.1 Compliance Audit Approach:

In this approach the audit team looks for deviations from the laws and company policies or procedures. As per the study conducted by Schwind, Das & Miner (1985), the fact-finding efforts of HR audit team in this approach try to find out whether there is compliance with the company policies and legal regulations. Similarly, Delery, J., & Gupta, N. (2016) stated on the review of past actions and its compliance with law and organisation policy. According to Clardy (2004), compliance audit can critically assess the effectiveness of control performed by the human resource management system and procedures. Hence, the main objective of the compliance audit approach is to provide principles and guidance for organizations that are designing, developing, implementing or improving a conformity program. The compliance audit covers legal requirement, industry and organizational standards and codes, the principles of good governance, code of conduct and ethical standards (Kaplan, 2008).

2.2 Productive Audit Approach:

The basic aim of this audit approach is to identify productivity and check whether employees are delivering according to organizational objectives and focuses on human resource contribution to the business strategy. The influence of an individual on productivity is the most difficult to measure in traditional cost accounting term (Lee, M. T., & Raschke, R. L., 2016). Productivity of most organization's is a function of at least three variables managed i.e. Technology, Capital and human resource and the influence of an organization HR audit on productivity, can be measured in term of what people do in the job; what people do can be appraised on the basis of attendance, timely achievement of targets, accident, turnover and amount of grievances; as well as what people do can be measured directly in term of observation by managers, peer, subordinate and customers as to the frequency with which employee do the things which are critical to job success (Beardwell, J., & Claydon, T., 2007). As per Edwards (2007), most organizations have not yet totally explored in a systematic manner the development of effective HR audit system. On theoretical perspective of HR audit, researchers are divided on the issue of best approach of carrying out the HR audit which is universally applicable on productivity (Andrews, 2007). Many organizations have been leaders in realizing the opportunity from technology development, however, have failed miserably to maximize the productivity by taking full advantage of the abilities of their employees by auditing human resource productivity.

2.3 Value for Money Audit Approach:

Performance/productivity audits seeks to assess the performance of the system or performance of employee as well as organisation. On the other hand, compliance audit is carried out to confirm that all rules and principles/procedures followed in conformity with the legal environment and organisation policy. However, these audits are not comprehensive enough, since they do not examine the economic aspect of human resources. As per Craven & McNulty (1994), an activity is worthwhile for an auditor, when it provides value-for-money to qualify as efficient, economic and effective. For instance, in the context of training expenditure, the expenditure is effective if it achieves its objectives without taking account of cost, it is economic if it achieves the most desired objectives at the least cost and it is efficient if it give the highest net present value when other uses of the fund are considered (Institute of internal auditors, 1986). Hence, employee's assessment for their value for money has become an important aspect of management and is one of the factors that have stimulated the spread of performance measurement (Wanyama, 2013).

3. Methodology:

In this research work descriptive survey research design has been followed for analysis of the objective. Basically, descriptive design is suitable for establishing the relationship between various events, processes or entities and is appropriate for this kind of study, since it will give flexibility in exploratory and descriptive data (Kothari, 2004; Mugenda and Mugenda, 2003). Stratified random sampling was adopted to select respondents from 04 numbers of organisations involved in electricity generation, transmission and distribution operating in Arunachal Pradesh. The population of this study was 418 consisting of 149 Managers and 269 operatives (Both technical and non-technical employees). Sample size of 146 which is 35% of population was determined as per Israel (2009) at 95% confidence level and respondents were picked up randomly from each stratum dependent on its proportion. As per Mugenda and Mugenda, (2003), a sample size 30% of population less than 500 is considered to be a representative sample. In order to achieve adequate distribution of the number of questionnaires for all the organisations in the study area, 150 questionnaires distributed for a sample size of 146, but only 128 responses were received i.e. the response rate is 85% and 120 responses were found valid on scrutiny and is used effectively for further statistical analysis, which is almost close to the determined sample size. Data were collected by administering questionnaire containing the following variables measured on summated five-point Likert scale.

3.1 Compliance Audit Approach: In Compliance audit approach the audit team looks for deviations from the laws and company policies or procedures. Through their facts finding effort, the team can determine whether there is compliance with the company policies and legal regulations. This variable was measured on a 5-item scale.

Table-2: Items for measuring compliance audit approach

No.	Items	Mean	Std. Dev.
1	Our organization follows rules and regulation on recruitment of employees.	4.4333	0.73030
2	Compensation and benefit laws are followed in our organization and salaries are paid in time.	4.4250	0.69406
3	Our organisation follows Health and safety policies.	4.4083	0.80436
4	In our organization promotions and transfers are conducted fairly without any bias and prejudice.	3.9000	1.16964
5	There are few or no cases of sexual harassment in our organization.	3.3667	1.14447

3.2 Productivity Audit Approach: Productivity audit deals with assessing the efficiency and effectiveness of a company's existing resources such as manpower in this case. This variable was measured on a 5-item scale.

Table-3: Items for measuring productivity audit approach

No	Items	Mean	Std. Dev.
1	The organisation follows all round productivity assessment of employees.	3.64	0.96837
2	The organisation has clear cut mechanism of observing employee productivity.	2.82	1.20212
3	Organisation has clear expectation in relation to productivity.	3.83	0.65251
4	Our services delivered to customers would improve if my production results are taken into consideration.	3.78	1.02459
5	There is always implementation of our productivity assessment.	3.73	0.96982

3.3 Value-for-Money Audit Approach: This approach of audit is generally adopted to describe a situation where an aspect of human resource is audited with a view to evaluating the performance on the basis of efficiency, effectiveness and economy. Unlike PAA and CAA this variable was measured on a 4-item scale.

Table-4: Items for measuring Value for Money audit approach

No.	Items	Mean	Std. Dev.
1	The organization always assess our value for money spent on us	3.7250	1.06875
2	Employees are considered as the most valued resource in the orgn.	3.6250	1.06165
3	The Organisation efficiently utilizes its human resource.	3.5000	.94380
4	There is always implementation of our value assessment.	3.1583	1.17392

3.3 Organisational Performance: In this study subjective measure of organisational performance was employed. The objective of utilising subjective measure as opposed to objective measure is because of few sampled organisations were not ready to disclose their financial performance. The subjective measure of performance was successfully used in CRANET (Tregaskis, Mahoney and Atterbury, 2004). In order to measure organizational performance, the scale developed by Bontis, Crossan and Hulland (2002) with the objective of collecting perceptions of business performance has been used along with some more items. The scale items are provided in the Table-5.

Table-5: Items for measuring Organisational Performance

No.	Items	Mean	Std. Dev.
1	The organisation serves its employees with all they need efficiently	4.06	0.946
2	My organisation is successful	3.32	1.195
3	The team meets its performance targets with well utilisation of resources without wastage	3.80	1.050
4	Individuals are generally happy working here	3.83	1.133
5	My organization's future performance is secure	4.04	0.844
6	The response time to customer needs is prompt	3.44	1.431
7	Employees positively contribute to the organization's growth	4.09	0.995

Finally, Ordinary Least Square (OLS) regression method adopted to test the impact of HR audit approaches on organisational performance. It is most widely used technique helpful in the examination of relationship between two or more variables. In the process of performing regression analysis, the result suggests which factors or variables matter most and which factors need to be ignored. A simple multiple regression equation can be written as follows:

$$Y = \alpha + \beta_1 X_1 + \beta_2 X_2 + \cdots + \beta_t X_t + u$$

Where, Y = Dependent Variable or endogenous variable (Organisation Performance)

X_i = Independent or Predictors or exogeneous variable (HR audit approaches) and

u = Error term; α and β_i are the coefficients of the equation.

4. Analysis and Discussion:

The regression analysis basically precedes determination of correlation among the variables. The result of correlation provided in Table-6 suggests that out of 03 audit approaches considered in this study the Compliance audit approach has highest significant positive correlation with organizational performance at 0.439. This may be a pointer that firms and organizations that emphasize on assessment of employment laws, rules and regulations and constantly audit adherence to employment related legal requirements will improve their performance, which supports the findings of Clardy (2004) "who found out that compliance audit can be used to assess effectiveness of control by critiquing human resource management systems and procedures". However, the correlation coefficient among productivity audit approach and value-for-money audit approach with organizational performance although positive and significant but are very loosely correlated. Moreover, all the explanatory variables were not strongly correlated signifying avoidance of multi collinearity while performing the regression analysis.

Table-6: Correlation of HR audit with organizational Performance

	OrgPer	CAA	VMA	PAA
OrgPer	1.000			
CAA	0.439**	1.000		
VMA	0.256	0.238**	1.000	
PAA	0.383**	0.278**	0.226*	1.000

The results in Table-7 shows that all the three audit approaches i.e. compliance audit, productivity audit and value for money audit were responsible in 34% variation in the organizational performance. The F-value (19.97) proves that the regression of HR audit approaches on firm performance was significant at $p < 0.05$, which was the evidence of goodness of fit of the regression model. Although, significant correlation among the exogeneous variables not warranted yet the collinearity diagnosis (Table-7) proves absence of multicollinearity, since the VIF values are within accepted limits. Similarly, the Durbin Watson test statistics signifies that residuals are relatively normal. The beta coefficient of compliance audit approach was significant at 5% level, while coefficient of productivity audit approach is significant at 10% level. Therefore, the regression equation based on these results can be expressed as follows:

$$Org.Perf = 1.556 + 0.462 * CAA + +0.122 * PAA + 0.014 * VMA$$

Table-7: Regression of HR audit approaches and Org. Performance

Model	R	R Square	MSE	F-Value	df_1	df_2	P-value
1	0.580ª	0.336	4.794	19.969	3	116	0.000 < 0.05
Model	Coeff.	Std. Error	t	Sig.		VIF*	Durbin Watson**
(Constant)	1.556	0.384	4.053	0.000			1.765
CAA	0.462	0.069	6.716	0.000		1.123	
PAA	0.122	0.073	1.662	0.099		1.116	
VMA	0.014	0.082	0.170	0.866		1.092	

a. Predictors: (Constant), Compliance Audit, Productivity Audit and Value-for-Money Audit approaches; Dependent Variable: Organization Performance

As per the equation above, the value of intercept term is 1.556, meaning that when other factors were zero, the organisational performance will have a value of 1.556. Moreover, the compliance audit accounted for 46.2% variation in the organisational performance. On the other hand, productivity audit is responsible for 12.2% and value-for-money audit 1.4% variation in organisational performance if considered in isolation respectively. While, the composite impact of predictors on the dependent variable is 59.8%.

5. Conclusion and Recommendations:

This paper has made an attempt to unearth the relationship between HR audit and organisation performance by adopting a quantitative method of analysis. Hence, this article at first reviewed and appraised the practice of HR auditing from extant literature. It also reviewed various approaches of HR audit and paid special attention to compliance, productivity and value-for-money audit approaches. The reason for focusing on government undertaking organisations is that the prominent feature of audit will shed light on various HR related issues and bring them into focus of the management. In order to analyze the impact of various HR audit approaches on organisation performance, correlation and OLS regression tool has been adopted.

The findings from correlation table provides highest positive correlation among compliance audit approach and organisational performance, that signifies higher emphasis on assessment of employment laws, rules and regulations. It is evident in all government undertaking. Thus, the constant audit adherence to employment related legal requirements will improve their performance, which supports the findings of Clardy (2004). From

the regression results it is proved that compliance audit approach is responsible highest variation in organisation performance in isolation. On the other hand, lower level of association was observed in productivity and value-for-money audit approaches with organisation performance. Moreover, in the regression analysis both these endogenous variables combinedly accountable of 13.6% variation in performance of organisation. This is an indication that although auditing the productivity of the human resource will improve performance of an organization, yet in the organisations under study importance of this variable is insignificant. The result contradicts the findings of the research on HR audit and organizational performance by Christopher Andrews (2007). Similarly, the correlation and regression coefficient of value-for-money audit approach and organizational performance is insignificant in the present context. This proves that organizations that audit the worth of their employees will not contribute to organizational performance. The finding contradicts the result of Aren (2005) which found that the workers value help to give an organization a competitive edge over others.
The findings and conclusions drawn recommends the government undertakings in the power sector who failed miserably in maximization of human resource productivity might take the full advantage of the abilities of their employees by auditing human resource productivity. In the same way, the economic aspect of human resource needs to be evaluated periodically, since it will stimulate the performance in the long run. Organisational factors like- management style, culture, technology, politics and size also influence organisation performance. Hence, further studies are required to find out how these organisational factors influence the relationship between HR audit and performance.

APPENDIX:

*A variance inflation factor (VIF) detects multicollinearity in regression analysis. Multicollinearity is when there's correlation between predictors (i.e. independent variables) in a model; it's presence can adversely affect your regression results. A rule of thumb for interpreting the variance inflation factor:1 = not correlated; between 1 and 5 = moderately correlated; Greater than 5 = highly correlated.
** The rule of thumb for DW test statistic values in the range of 1.5 to 2.5 are relatively normal. Values outside of this range could be a cause for concern (Field, 2009).

References

Beardwell, J., & Claydon, T. (Eds.). (2007). Human resource management: A contemporary approach. Pearson Education.Strunk, W., Jr., & White, E. B. (1979). *The elements of style* (3rd ed.). New York: MacMillan.

Bloom, N., & Van Reenen, J. (2006). Management Practices, Work-Life Balance, and Productivity: A Review of Some Recent Evidence. Oxford review of economic policy, 22(4), 457-482.

Burrowes, A., & Persson, M. (2000). The Swedish management audit: a precedent for performance and value for money audits. Managerial Auditing Journal.

Carter, C., Spence, C., Muzio, D., Andon, P., Free, C., & O'Dwyer, B. (2015). Annexing new audit spaces: challenges and adaptations. Accounting, Auditing & Accountability Journal.

Clardy, A. (2004). Toward an HRD auditing protocol: Assessing HRD risk management practices. Human Resource Development Review, 3, 124–150.

Delery, J., & Gupta, N. (2016). Human resource management practices and organizational effectiveness: internal fit matters. Journal of Organizational Effectiveness: People and Performance.

Edwards, J. E., Scott, J. C., & Raju, N. S. (2007). Evaluating human resources programs: A 6-phase approach for optimizing performance. John Wiley & Sons.

Field, A. (2009). Discovering statistics using SPSS:(and sex and drugs and rock'n'roll). Sage.

Gilligan, J. M. (2018). Accounting for the human factor. Nature Climate Change, 8(1), 14-15.

Hana, T. A. (2016). Human resource audit: the case of st. Mary's university. Ph.D. dissertation, St. Mary's University.

Kalubanga, M., & Kakwezi, P. (2013). Value for money auditing and audit evidence from a procurement perspective-A conceptual paper. International Journal of Advances in Management and Economics, 2(5), 115-124.

Kaplan, R. S., & Norton, D. P. (2008). The execution premium: Linking strategy to operations for competitive advantage. Harvard Business Press.

Kothari, C. R. (2004). Research Methods, Quantitative and Qualitative Approaches.

Lee, M. T., & Raschke, R. L. (2016). Understanding employee motivation and organizational performance: Arguments for a set-theoretic approach. Journal of Innovation & Knowledge, 1(3), 162-169.

McConnell, J. H. (2011). Auditing Your Human Resources Department: A Step-by-step Guide to Assessing the Key Areas of Your Program. Amacom Books.

Mugenda, O. M., & Mugenda, A. G. (2003). Research methods: Quantitative and qualitative approaches. Acts press.

Nutley, S. (2000). Beyond systems: HRM audits in the public sector. Human Resource Management Journal, 10, 21.

Olalla, M. F., & Castillo, M. A. (2002). Human resources audit. International Advances in Economic Research, 8, 58–64.

Parker, L. D., Ferris, K. R., & Otley, D. T. (1989). Accounting for the human factor. Prentice Hall.

Rangriz, M. A. (2014). The human resources auditing. Journal of Management Value & Ethics.

Rao, T. V. (2014). HRD audit: Evaluating the human resource function for business improvement. SAGE Publications India.

Schuler, R. S., Budhwar, P. S., & Florkowski, G. W. (2002). International human resource management: review and critique. International Journal of Management Reviews, 4(1), 41-70.

Schwind, H., Das, H., & Miner, F. (1985). Canadian Personnel Management and Human Resources.

Shaban, O. (2012). Auditing human resources as a method to evaluate the efficiency of human resources functions and to control quality check on HR activities. International Business Research, 5(3), 122.

Ukil, M. I. (2015). Essence of Human Resource Audit: An Analytical Study. Journal of Management and Science, 5, 101–110.

Yadav, R. K., & Dabhade, N. (2014). Human resource planning and audit–a case study of HEG limited. human resource management, 4, 5.

Effect of Covid-19 pandemic on the Hospitality sector in India with specific reference to Itanagar, Arunachal Pradesh.

K Mariam[1], M M Singh[2] and T Tadhe Goyang[3]

[1,2,3] National Institute of Technology, Arunachal Pradesh-791112, India

kagomariam@gmail.com

Abstract. Covid-19 has left many businesses and is still affecting many businesses in the world. The ripple effect can be seen throughout mankind and business entities causing huge loss of life as well as jobs. Hospitality sector is worst hit by pandemic. There is a major loss to business during this pandemic with people staying homes and not venturing out leading to the worst loss of businesses in the history. Huge cancellations of tours, events, flights, hotel reservation, restriction in inbound travel, suspension of PAP (Protected Area Permit) and ILP (Inner Line Permit) have led to low hotel occupancy percentage as well as the average room rates have fallen severely leading to extraordinary losses. Also, travel advisories issued by the World Health Organization to discourage travel, government movement restriction orders, lockdown orders and campaign of stay home stay safe to curb the spread of virus have added to the burden of the hospitality sector.Hospitality industry will not be the same as before during covid-19 even if lockdowns and Government order on ban of movement & travel will be eased. But the hospitality service providers cannot afford to remain shut down till the vaccine is found or operate with the existing old model of operations. Individual crisis management plans along with the guidelines set by the govt. and many important measures and innovations need to be adopted by the hotels in order to gain customer's confidence, survive and thrive. Some hopeful govt. supports to help stabilize hotel business can be relief funds, tax cuts, subsidies, relaxation on license renewal, credit facilities, employment support, vocal for local etc.

1. Introduction

Prior to covid-19 most of the countries in the world have accepted tourism & hospitality as the major contributor to economic growth. Even in India tourism & hospitality is considered as the positive contributor towards economic growth and employment generation. It is evident by various government schemes and policies in favor of tourism & hospitality to attract more tourists in order to gain more revenue such as Atithidevobhava, Incredible India and various state tourism campaigns, projects and investments. Crores of funds are generated & invested to develop a tourist destination in order to attract huge tourist arrival and in return earn revenues. Government had strict guidelines as well as extreme care was undertaken by various stakeholders in giving a great tourist experience in order to maintain the reputation as well as to retain the hospitality business. However whenever there has been any major impact on the economy of any country, whether it is political unrest, terrorism, pandemic, natural calamity etc. tourism & hospitality has always been the first one to get affected as people start to cut down on their travel plans and budget. The tourism industry since it involves movement & travel of people from one corner to another corner of the world, in a way it has carried the burden of covid 19 pandemic severely. Travel has been blamed as carriers as well as the sufferer of the novel corona virus. To stop the transmission of corona, the respected Indian government has strictly stopped the movement of people which has tremendously affected the normal travel business. India has witnessed adverse effects of corona in terms of cancellation of hotel bookings; pull back of tour programme, closed down of restaurants, postponement of events etc. All these have incurred huge losses to the normal business of hospitality sectors in India as well as losses of employment.

Due to COVID-19 we have a new normal that includes social distancing, avoiding unnecessary outings, health protocols, safety, sanitation and hygiene. Post covid 19 customers or hotel guests will look out more for their wellbeing, security and healthy environment rather than luxury or great services. Therefore the first step towards having normal hotel business is ensuring standard operating procedures amid covid-19 [Tse et al., 2006] and minimizing unnecessary operating expenses in order to have long term sustainability.

2. Literature survey

Arunachal is the eastern most state in India, with Itanagar as its capital city. We have witnessed a huge blow to our state hospitality sector due to corona whether it is curfew, closure of business premises, shut down of school and colleges, lockdowns, work from home orders, inter district and interstate movements restrictions, temporary suspension of PAP and ILP etc. These challenges have resulted in temporary closure of many hospitality businesses and thereby threatened the occupation and survival of many service providers and their employees. Huge revenue loss due to closure of various tourism services was witnessed. To control the spread of coronavirus tourist traffic was strictly restricted which led to abrupt decline in tourist arrivals which directly impacted the flow of revenue due to no tourism activity [Thashneem T. et al., 2020].

Hospitality industry consists of various sectors and subsidiary services. Hotels should cooperate and perform with other stakeholders to prepare the rescue tactics to regain its foothold considering the fact that it depends on such subsidiary services for its survival. Decrease in demand for hotel rooms and other hotel businesses like events and functions has led to decline in ARR (average room rate) and REVPAR (revenue per available rooms). Hence the hotels need to plan firstly on cost-cutting measures to sustainably run on long terms. Services and areas with very less or no request should be identified and for a time being till the business stabilizes, be put under temporary closure measures in order to curb wastage. Unnecessary working equipment and heavy machinery, heavy cost incurred areas can be shut down or have strict controlled supervision so as to optimise the expenditure. Some examples are shutting down sections of guest rooms, proper planning on assigning of rooms that can be occupied or laid down, restaurant planning, controlled use of guest elevator, air conditioning planning especially in case of centralized system, lightings, controlled hotel laundry services etc [Mensah,2020].

Apart from mentioning the above mentioned measures there is a very important area which cannot be neglected in spite of cost involved, that is training and re-training. Training can be imparted to the employees on multitasking various duties & responsibilities to survive the lean season of the hotel business [Kyriakidou and Maroudas, 2010]. Reduced workloads and lesser responsibilities because of low occupancy in the hotel added with fewer manpower makes it inevitable to learn to multitask. Examples of some training that can be imparted are on hygiene, wellbeing, stress control, healthy lifestyle, new SOP training programs, departmental training programs etc. It is to ensure efficiency, professionalism, quality services as well as to improve the morale of your staff. Your employees are your strength and brand ambassador of your hotel. Trainers can be hired from within the organizations or from outside. These training can be conducted on the job or virtually using various online modes or information and communication tools (ICT). Employees should be familiarized with new protocols such as wearing masks or face shields as part of uniforms, use of sanitizers, frequent cleaning of surfaces, frequent washing of hands and wearing of PPEs in case of emergency etc. The hotel should be installed with hygiene equipment such as sanitizers, special air filters, procurement of thermal scanners. New innovations in service industry such as air purifier, disinfectant diffusers, whole body scanner, use of robots for cleanliness and sanitation purposes especially in back of the house areas. Technology such as automated systems and digital systems that can be used for hotel day to day business purposes such as check ins, reservations, functions etc. can be adopted [Webster and Ivanov.,2020; Yang et al., 2020]. The use of robots to clean and fumigate guestrooms and back areas of the premises as well as other public areas can aid in preventing the virus contamination through people. This adoption of new innovation where there is less human interaction but more service efficiency can aid in building customers confidence and in attracting more customers to try out the new normal. However the hospitality and service industry can never run without human touch but this is the new normal due to covid-19 pandemic and we need to learn to survive in such situations. Hotels and hospitality service providers must adhere to strict and disciplined SOP [VKaushal and S Srivastava., 2020]. Any reported instance of existence or news of the virus in a certain hotel or service provider will extremely and negatively tarnish the reputation of that specific service provider.

3. Objective of the study

i) To study the impact of covid-19 on the hospitality sector in Itanagar, Arunachal Pradesh.
ii) To bring out possible solutions and measures in controlling the impact of covid-19 in the hospitality sector in itanagar, Arunachal Pradesh.

4. Methodology

The study is based on primary data collected through e questionnaire and telephonic interview. Secondary data available such as magazines, journal publications, books, reports, internet etc. were also referred and used. The findings are discussed on the purview of the available resources.

5. Results and Discussion

5.1. Influence of Corona on tourist arrival in the state of Arunachal Pradesh.

Tourist arrivals in Arunachal Pradesh for the last five years were taken as shown in table 1. Tourism and hospitality business was performing very decent prior to covid-19. There is a positive growth of 57.77% on domestic tourist arrival and 21% on foreign tourist arrival. However, had there been no covid-19 pandemic, forecasted tourist arrival based on time series analysis for the year 2020 and 2021; domestic tourist arrival for the year 2020 is showing 609913.5 and for the year 2021 is showing 663262.2 in the Figure 1 below. Similarly forecasted foreign tourist arrival for the year 2020 is showing 8274.9 and 2021 is 8654.8 in the Figure 2 below.

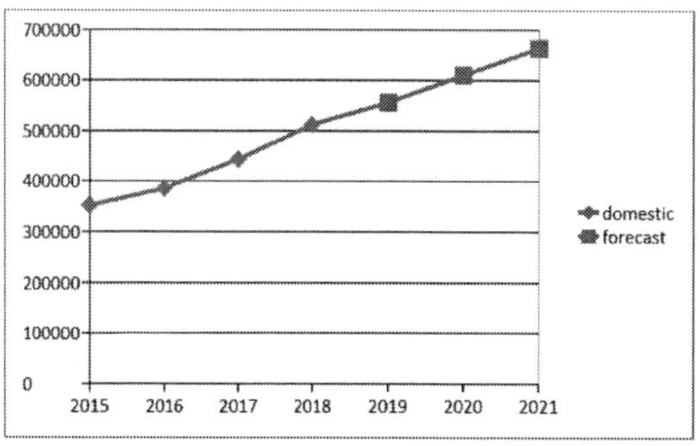

Figure 1. Forecasted foreign domestic arrival for the yr. 2020 and 2021

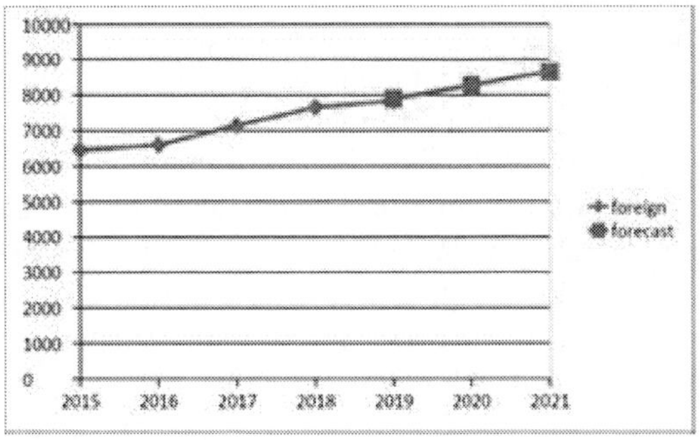

Figure 2. Forecasted foreign tourist arrival for the yr. 2020 and 2021

Table 1. Forecasted domestic tourist arrival for the yr. 2020 and 2021

Year	Domestic Tourist Arrival	Foreign Tourist Arrival	
2015	352176	6453	-
2016	385875	6598	-
2017	443211	7147	-
2018	512436	7653	-
2019	555639	7825	-
2020	609913.5	8274.9	Forecasted
2021	663262.2	8654.8	Forecasted

(source: Office of Directorate of Tourism, Dept. of Tourism, Itanagar, Arunachal Pradesh).

5.2. Influence of Corona on Hotels of Itanagar, Arunachal Pradesh.

Primary data was collected using e questionnaire and telephonic interview to understand the impact of covid 19 pandemic on the hospitality sector in Itanagar, Arunachal Pradesh India. Itanagar roughly has 29 registered hotels with lodging facilities. Prior to covid-19 things were normal for these hospitality service providers. Based on a random sample, 9 hotels were surveyed. The lockdown has had an unfortunate impact on the hospitality industry. It has resulted in a very low occupancy percentage consequently leading in some bigger hotels having to either shut operations or run with very minimum low facility. It is to be noted that most of the hotels currently have occupancy of not more than 5% to 12% and these are the people who are on quarantine or front line workers as per govt. order. Interestingly small hotels or budget hotels were least affected in terms of occupancy and having almost occupancy percentages of 60% to 80%. Similarly normal hotel operations such as guest arrival, reservation, pre booking, events and functions were severely affected in larger hotels whereas budget hotels or smaller hotels were not affected much. There is a mixed response towards the salary cut of regular staff and laying off staff. However budget and smaller hotels again seemed to do better in this category. Some hotels even stated to be understaffed since many of the staff had left for their hometown to check upon their family members that reside in other neighbouring states such as Assam but could not return back on the job due to temporary closure of ILP services in the state to curb corona transmission. Reasons behind smaller hotels doing alright could be credited to state government order for quarantine of the returnees from other states and accommodating front line workers in the hotels of the capital complex. With fewer rooms & fewer facilities provided in the smaller hotels there are lesser expenditure on the fixed cost and other operating cost such as electricity bill, water bill, payroll, govt taxes, license fee, guest room amenities, linen & toiletries, food & beverage inventories etc, which turned helpful in surviving the lockdown period announced by the state Government.

6. Findings

The state Government order that stated the use of hotel rooms for quarantine purpose for the returnees of residents of Arunachal Pradesh from other states as well as for the use of hotel rooms by front line covid warriors; has definitely benefited the budget or small hotels in the larger interest during this pandemic. It helped small hotels with fewer rooms and facilities to cover their fixed cost as well as other operating costs efficiently. Which otherwise has severely affected the larger hotels in terms of revenue earning and survival.

All the hotels agreed upon the notion that standard operating procedures SOP have to be strictly and diligently followed upon opening of hotel business which however will pick its pace by early next year or unless vaccine is found. They have all uniformly stated that they seek various supports from government for helping them stabilizing the hotel business such as relaxation in tax and deadline of paying taxes, free extension of various licenses for running of business, sanitation supplies, subsidies, relief funds, discount in payment of utility bills, employment support, credit facilities, loans with smaller interest etc.

7. Conclusions

The damage caused by this pandemic is massive but surely it will settle down sooner or later. The world will learn to stay normal and simultaneously with the pandemic. We need to remain ready when we will achieve normalcy (COVID-19-Impact-on-the-Indian-Hotels, 2020). It has taught us one thing that we cannot depend on the inbound tourism market as in the present scenario cases. Hence for the hospitality sector to thrive, it must approach the local or native people and domestic market by having some interesting, simple, innovative and attractive promotional packages. Local business and people involved in the hospitality sectors need to be encouraged for gainful employment. We need to be vocal for local initiatives by our state government. One of the ways could be by these hotels getting in touch with the local farmers for organic produce and direct purchase without involving middlemen. This could be also a gesture towards corporate social responsibility (CSR). It will benefit both the party that is the hotel and the local people. Special packages and value add focusing on the 'locals' should be the new focus. Hotels can give services that were previously not part of the hotel's services to guests such as online food delivery or letting off hotel kitchens for cloud kitchen systems etc. Government & the hospitality management in collaboration can also help attract customers and encourage them to visit hotels or use its various services by giving discounts and attractive offers initially to support and encourage customers in saving the business. Hotels need to provide suitable assistance to the returning employees to once again resume their professional life with proper training etc. including aid on personal matters such as housing, medical etc. Government needs to encourage hotels to hire local people for the job by giving incentives and employment support. Hotels need to ensure that they sincerely and strictly follow their own internal guidelines as well as the SOP guidelines given by the government.

References

[1] Lama,M.(2020,July05).Arunachal loses 2.02 lakh tourist arrivals due to Covid-19 lockdown.The Arunachal Times. https://arunachaltimes.in/index.php/2020/07/05/

arunachal-loses-2-02-lakh-tourist-arrivals-due-to-covid_19lockdown
[2] Kumar,D.(2020,May06).Handling The Covid-19 Crisis & Comeback Planning - A Way Forward.BW Hotelier. http://bwhotelier.businessworld.in/article/Handling-The-Covid-19-Crisis-Comeback-Planning-A-Way-Forward
[3] Coronavirus-Arunachal Pradesh bans entry of foreign tourists(2020,March08).The Hindu. https://www.thehindu.com/news/national/arunachal-bans-entry-of-foreign-tourists/article31016616.ece
[4] Nusra.(2020,March26).Covid-19 and its Impact on Hotel & Tourism Industry.Restaurant India.https://www.restaurantindia.in/article/covid-19-and-its-impact-on-hotel-tourism-industry.13612
[5] Kishore,R.,Jha,A.(2020,June14).Economic impact of Covid-19 pandemic to vary in sectors.The Hindustan Times. https://www.hindustantimes.com/india-news/economic-impact-of-covid-19 -pandemic-to-vary-in-sectors/story-DIWjwnBZoON7ZUvgSMSFOL.html
[6] Jain.V,Jain,P.(2020,April25).Impact of covid-19 on Independent/Budget Hotel Industry in India.http://bwhotelier.businessworld.in/article/Impact-of-Covid-19-on-Independent-Budget-Hotel-Industry-in-India/25-04-2020-190282/
[7] COVID-19: Impact on the Indian Hotels Sector, A Report by HVS.(2020,April10).BwHotelier. http://bwhotelier.businessworld.in/article/COVID-19-Impact-on-the-Indian-Hotels-Sector-A-Report-by-HVS/10-04-2020-188770/
[8] Covid19 hits hotel and hospitality sector; Delhi, Bengaluru worst-hit.(2020,May,13).Deccan Chronicle https://www.deccanchronicle.com/business/in-other-news/130520/covid19-impacts-hotel-and-hospitality-sector-delhi-bengaluru-worst-h.html
[9] Talukdar,B(2020,May,19).COVID-19 & tourism in Northeast post 2020 .https://www.sentinelassam .com / editorial/covid-19-tourism-in-northeast-post-2020-477561
[10] Vikrant Kaushal.,SidharthSrivastava. Hospitality and tourism industry amid COVID-19 pandemic: Perspectives on challenges and learnings from India.Int J HospManag. 2021 Jan; 92: 102707. Published online 2020 Oct 1. doi: 10.1016/j.ijhm.2020.102707
[11] Kyriakidou O., Maroudas L. Training and development in British hospitality, tourism and leisure SMEs. Manag.Leis. 2010;15(1-2):32–47. doi: 10.1080/13606710903447998.
[12] Tse A.C.B., S.S., Sin L. Crisis management and recovery: how restaurants in Hong Kong responded to SARS. Int. J. Hosp. Manag. 2006;25(1):3–11. doi: 10.1016/j.ijhm.2004.12.001.
[13] Webster C., Ivanov S. Demographic change as a driver for tourism automation. J. Tour. Futures. 2020 doi: 10.1108/JTF-10-2019-0109.
[14] Thashneem T. Bhanu., Prasanna Kumar J.P. The Effect of COVID-19 Lockdown on Indian Tourism and Hospitality Industry.Journal of Dental & Oro-facial Research Vol. 16 Issue 02 Aug. 2020 JDOR 34.
[15] L. Yang, T.L. Henthorne, B. George. Artificial intelligence and robotics technology in the hospitality industry: current applications and future trends. Digital Transformation in Business and Society, Palgrave Macmillan, Cham (2020), pp. 211-228
[16] Andrews,S(2020,April,18).Impact of COVID-19 on the hotel industry and suggestions for post-pandemic recovery. https://hospitality.economictimes.indiatimes.com/blog/impact-of-covid-19-on-the-hotel-industry-and-suggestions-for-post-pandemic-recovery/4214
[17] Mensah,I.(2020,May13).Hospitality Net.https://www.hospitalitynet.org/opinion/4098657.html

APPROXIMATE SERIES SOLUTION OF NONLINEAR INHOMOGENEOUS TIME FRACTIONAL PARTIAL DIFFERENTIAL EQUATION USING GENERALIZED DIFFERENTIAL TRANSFORM METHOD

Deepanjan Das[1], Madan Mohan Dixit[2] and Chandra Prakash Pandey[3]
[1]Department of Mathematics, Ghani Khan Choudhury Institute of Engineering and Technology, Narayanpur, Malda,West Bengal-732141, India.
[2]Department of Mathematics, North Eastern Regional Institute of Science and Technology, Nirjuli, Itanagar, Arunachal Pradesh-791109, India.
[3]Department of Mathematics, North Eastern Regional Institute of Science and Technology, Nirjuli, Itanagar, Arunachal Pradesh-791109, India.

E-mail:gkcietdeepanjan@gmail.com

Abstract. In the present paper, we obtain the approximate solution of nonlinear inhomogeneous time-fractional partial differential equation by Generalized Differential Transform Method (GDTM). The fractional derivatives are described in the Caputo sense. GDTM provides an analytical solution in the form of an infinite power series with easily computable components. GDTM can solve the linear and nonlinear fractional differential equations with negligible error compared to the exact solution. The method introduces a promising tool for solving many linear and nonlinear fractional differential equations.

Keywords: Fractional differential equations; Caputo fractional derivative; Generalized Differential transform method; Generalized Taylor formula; Analytic solution.

Mathematical Subject Classification (2010): 26A33, 34A08, 35A22, 35K25, 35R11, 74H10.

1. Introduction:

In the present analysis GDTM have been used to solve the following nonlinear inhomogeneous time-fractional partial differential equation

$$\frac{{}^{c}\partial^{\alpha}u(x,t)}{\partial t^{\alpha}}+x\frac{\partial u(x,t)}{\partial x}+\frac{\partial^{2}u(x,t)}{\partial x^{2}}+cu^{2}(x,t)=2t^{\alpha}+2x^{2}+2\ ;\quad t>0\,,$$

subject to initial condition $u(x,0)=x^{2}$, (1)

where $\frac{\partial^{\alpha}}{\partial t^{\alpha}}$ is the fractional differential operator(Caputo derivative) of order $\alpha\in(0,1]$ and c =constant. Fractional differential equations are generalizations of classical differential equations of integer order and have recently been proved to be valuable tools in the modeling of many physical phenomena in various fields of science and engineering. A lot of works have been done by using fractional derivatives for a better description of considered material properties. Mathematical modeling based on enhanced rheological models naturally leads to differential equations of fractional order and to the necessity of the formulation of the initial conditions to such equations.

Recently, various analytical and numerical methods have been employed to solve linear and nonlinear fractional differential equations. The differential transform method (DTM) was proposed by Zhou [1] to solve linear and nonlinear initial value problems in electric circuit analysis. This method has been used for solving various types of equations by many authors [2-16] . DTM constructs an analytical solution in the form of a polynomial and different from the traditional higher order Taylor series method. For solving two-dimensional linear and nonlinear partial differential equations of fractional order DTM is further developed as Generalized Differential Transform Method (GDTM) by Momani, Odibat, and Erturk in their papers [17-19].

Recently, Vedat Suat Ertiirka and Shaher Momanib applied generalized differential transform method to solve fractional integro-differential equations [20]. The GDTM is implemented to derive the solution of space-time fractional telegraph equation by Mridula Garg, Pratibha Manohar and Shyam L.Kalla [21]. Manish Kumar Bansal and Rashmi Jain applied generalized differential transform method to solve fractional order Riccati differential equation [22].

2. Mathematical Preliminaries on Fractional Calculus

We introduce the following definitions [23, 24] in the present analysis:

2.1. Definition: Let $\alpha \in \boldsymbol{R}^+$.The integral operator I^α defined on the usual Lebesgue space L (a, b) by

$$I^\alpha f(x) = \frac{d^{-\alpha} f(x)}{dx^{-\alpha}} = \frac{1}{\Gamma(\alpha)} \int_0^x (x-t)^{\alpha-1} f(t)dt$$

and $I^0 f(x) = f(x)$,
for $x \in [a,b]$ is called Riemann-Liouville fractional integral operator of order α (≥ 0).
It has the following properties:
(i) $I^\alpha f(x)$exists for any $x \epsilon [a,b]$,
$(ii) I^\alpha I^\beta f(x) = I^{\alpha+\beta} f(x)$,
(iii) $I^\alpha I^\beta f(x) = I^\beta I^\alpha f(x)$,
(iv) $I^\alpha x^\gamma = \frac{\Gamma(\gamma+1)}{\Gamma(\alpha+\gamma+1)} x^{\alpha+\gamma}$,
where $f(x) \epsilon L[a,b], \alpha, \beta \geq 0, \gamma > -1$.

2.2. Definition: The Riemann-Liouville definition of fractional order derivative is
$${}^{RL}_{0}D^\alpha_x f(x) = \frac{d^n}{dx^n} {}_0I^{n-\alpha}_x f(x) = \frac{1}{\Gamma(n-\alpha)} \frac{d^n}{dx^n} \int_0^x (x-t)^{n-\alpha-1} f(t)\,dt\,,$$
where n is an integer that satisfies $\alpha \in (n-1, n)$.

2.3. Definition: A modified fractional differential operator ${}^C_0D^\alpha_x$ proposed by Caputo is given by
$${}^C_0D^\alpha_x f(x) = {}_0I^{n-\alpha}_x \frac{d^n}{dx^n} f(x) = \frac{1}{\Gamma(n-\alpha)} \int_0^x (x-t)^{n-\alpha-1} f^{(n)}(t)\,dt,$$
where $\alpha \in \boldsymbol{R}^+$ is the order of operation and n is an integer that satisfies $\alpha \in (n-1, n)$.

It has the following two basic properties [25]:
(i) If $f \in L_\infty(a,b)$ or $f \in C[a,b]$ and $\alpha > 0$, then ${}^C_0D^\alpha_x\ {}_0I^\alpha_x f(x) = f(x)$.
(ii) If $f \in C^n[a,b]$ and if $\alpha > 0$, then ${}_0I^\alpha_x\, {}^C_0D^\alpha_x f(x) = f(x) - \sum_{k=0}^{n-1} \frac{f^{(k)}(0^+)}{k!} x^k$; $\alpha \in (n-1, n)$.

2.4. Definition: For m being the smallest integer that exceeds α, the Caputo time-fractional derivative operator of order $\alpha > 0$, is defined as in [26]
$$D^\alpha_t u(x,t) = \frac{\partial^\alpha u(x,t)}{\partial t^\alpha} = \begin{cases} \frac{\partial^m u(x,\xi)}{\partial \xi^m}\ ; & \alpha = m \epsilon \mathbb{N} \\ \frac{1}{\Gamma(m-\alpha)} \int_0^t (t-\xi)^{m-\alpha-1} \frac{\partial^m u(x,\xi)}{\partial \xi^m} d\xi\ ; & \mathrm{m}-1 \leq \alpha < m \end{cases}.$$

Relation between Caputo derivative and Riemann-Liouville derivative:
$${}^C_0D^\alpha_t f(x) = {}^{RL}_{0}D^\alpha_t f(x) - \sum_{k=0}^{m-1} \frac{f^{(k)}(0^+)}{\Gamma(k-\alpha+1)} x^{k-\alpha}\ ;\ \alpha \in (m-1, m)\,.$$

Integrating by parts, we get the following formulae as given in [27]:

(i) $\int_a^b g(x)\, {}_a^CD_x^\alpha f(x)dx = \int_a^b f(x)\, {}_x^{RL}D_b^\alpha g(x)dx + \sum_{j=0}^{n-1}\left[{}_x^{RL}D_b^{\alpha+j-n} g(x)\, {}_x^{RL}D_b^{n-1-j} f(x)\right]_a^b$.

(ii) For $n = 1$, $\int_a^b g(x)\, {}_a^CD_x^\alpha f(x)dx = \int_a^b f(x)\, {}_x^{RL}D_b^\alpha g(x)dx + \left[{}_xI_b^{1-\alpha} g(x).f(x)\right]_a^b$.

3. Generalized two dimensional differential transform method

Consider a function of two variables $u(x,y)$be a product of two single-variable functions, i.e. $u(x,y) = f(x)g(y)$, which is analytic and differentiated continuously with respect to x and y in the domain of interest. Then the generalized two-dimensional differential transform of the function $u(x,y)$is given by [17-19]

$$U_{\alpha,\beta}(k,h) = \frac{1}{\Gamma(\alpha k+1)\Gamma(\beta h+1)}\left[\left(D_{x_0}^\alpha\right)^k \left(D_{y_0}^\beta\right)^h u(x,y)\right]_{(x_0,y_0)}, \tag{2}$$

where $0 < \alpha, \beta \leq 1$; $U_{\alpha,\beta}(k,h) = F_\alpha(k)G_\beta(h)$ is called the spectrum of $u(x,y)$ and $\left(D_{x_0}^\alpha\right)^k = D_{x_0}^\alpha, D_{x_0}^\alpha, \dots\dots, D_{x_0}^\alpha$ (k-times).

The inverse generalized differential transform of $U_{\alpha,\beta}(k,h)$ is given by

$$u(x,y) = \sum_{k=0}^{\infty}\sum_{h=0}^{\infty} U_{\alpha,\beta}(k,h)(x-x_0)^{k\alpha}(y-y_0)^{h\beta}. \tag{3}$$

It has the following properties:

(i) If $u(x,y) = v(x,y) \pm w(x,y)$, then $U_{\alpha,\beta}(k,h) = V_{\alpha,\beta}(k,h) \pm W_{\alpha,\beta}(k,h)$.

(ii) If $(x,y) = av(x,y)$,$a \in \mathbb{R}$, then $U_{\alpha,\beta}(k,h) = aV_{\alpha,\beta}(k,h)$.

(iii) If $(x,y) = v(x,y)w(x,y)$, then $U_{\alpha,\beta}(k,h) = \sum_{r=0}^{k}\sum_{s=0}^{h} V_{\alpha,\beta}(r,h-s)\, W_{\alpha,\beta}(k-r,s)$.

(iv) If $u(x,y) = (x-x_0)^{n\alpha}(y-y_0)^{m\beta}$, then $U_{\alpha,\beta}(k,h) = \delta(k-n)\delta(h-m)$.

(v) If $(x,y) = D_{x_0}^\alpha v(x,y), 0 < \alpha \leq 1$, then $U_{\alpha,\beta}(k,h) = \frac{\Gamma(\alpha(k+1)+1)}{\Gamma(\alpha k+1)} V_{\alpha,\beta}(k+1,h)$.

(vi) If $(x,y) = D_{x_0}^\gamma v(x,y), 0 < \gamma \leq 1$, then $U_{\alpha,\beta}(k,h) = \frac{\Gamma(\alpha k+\gamma+1)}{\Gamma(\alpha k+1)} V_{\alpha,\beta}\left(k+\frac{\gamma}{\alpha},h\right)$.

(vii) If $(x,y) = D_{y_0}^\gamma v(x,y), 0 < \gamma \leq 1$, then $U_{\alpha,\beta}(k,h) = \frac{\Gamma(\beta h+\gamma+1)}{\Gamma(\beta h+1)} V_{\alpha,\beta}\left(k,h+\frac{\gamma}{\beta}\right)$.

Where $U_{\alpha,\beta}(k,h), V_{\alpha,\beta}(k,h)$ and $W_{\alpha,\beta}(k,h)$are the differential transformations of the functions $u(x,y), v(x,y)$and $w(x,y)$ respectively and

$$\delta(k-n) = \begin{cases} 1 & ; \ k = n \\ 0 & ; \ k \neq n \end{cases}.$$

Applying generalized two-dimensional differential transform (2) with $(x_0,t_0) = (0,0)$ on (1) we obtain

$$U_{1,\alpha}(k,h+1)=\frac{\Gamma(\alpha h+1)}{\Gamma(\alpha(h+1)+1)}\Big[-\sum_{r=0}^{k}\sum_{s=0}^{h}\delta(r-1)\delta(h-s)(k-r+1)\,U_{1,\alpha}(k-r+1,s)-$$
$$c\sum_{r=0}^{k}\sum_{s=0}^{h}U_{1,\alpha}(r,h-s)\,U_{1,\alpha}(k-r,s)-(k+1)(k+2)U_{1,\alpha}(k+2,h)+2\delta(h-1)\delta(k)+$$
$$2\delta(h)\delta(k-2)+2\delta(h)\delta(k)\Big] \quad (4)$$

and $U_{1,\alpha}(k,0)=\delta(k-2)=\begin{cases}1 & ;\ k=2\\ 0 & ;\ k\neq 2\end{cases}$ (5)

Taking $p=h+1$ in (4) we get

$$U_{1,\alpha}(k,p)=\frac{\Gamma(\alpha(p-1)+1)}{\Gamma(\alpha p+1)}\Big[-\sum_{r=0}^{k}\sum_{s=0}^{p-1}\delta(r-1)\delta(p-s-1)(k-r+1)\,U_{1,\alpha}(k-r+1,s)-$$
$$c\sum_{r=0}^{k}\sum_{s=0}^{p-1}U_{1,\alpha}(r,p-s-1)\,U_{1,\alpha}(k-r,s)-(k+1)(k+2)U_{1,\alpha}(k+2,p-1)+2\delta(p-2)\delta(k)+$$
$$2\delta(p-1)\delta(k-2)+2\delta(p-1)\delta(k)\Big]$$

i.e. $U_{1,\alpha}(k,h)=\frac{\Gamma(\alpha(h-1)+1)}{\Gamma(\alpha h+1)}\Big[-\sum_{r=0}^{k}\sum_{s=0}^{h-1}\delta(r-1)\delta(h-s-1)(k-r+1)\,U_{1,\alpha}(k-r+1,s)-$
$$c\sum_{r=0}^{k}\sum_{s=0}^{h-1}U_{1,\alpha}(r,h-s-1)\,U_{1,\alpha}(k-r,s)-(k+1)(k+2)U_{1,\alpha}(k+2,h-1)+2\delta(h-2)\delta(k)+$$
$$2\delta(h-1)\delta(k-2)+2\delta(h-1)\delta(k)\Big] \quad (6)$$

Now utilizing the recurrence relation (6) and the initial condition (5), we obtain after a little simplification the following values of $U_{1,\alpha}(k,h)$ for $k=0,1,2,\ldots$ and $h=0,1,2,\ldots$
$U_{1,\alpha}(2,0)=1$; $U_{1,\alpha}(k,0)=0\ for\ k\in\mathbb{W}-\{2\}$;

$U_{1,\alpha}(0,1)=0$; $U_{1,\alpha}(0,2)=\frac{2\Gamma(\alpha+1)}{\Gamma(2\alpha+1)}$; $U_{1,\alpha}(0,3)=0$; $U_{1,\alpha}(0,4)=0$; $U_{1,\alpha}(0,5)=-\frac{4c(\Gamma(\alpha+1))^2\Gamma(4\alpha+1)}{(\Gamma(2\alpha+1))^2\Gamma(5\alpha+1)}$;

$U_{1,\alpha}(0,6)=0$; $U_{1,\alpha}(0,7)=0$; $U_{1,\alpha}(0,8)=\frac{16c^2(\Gamma(\alpha+1))^3\Gamma(4\alpha+1)\Gamma(7\alpha+1)}{(\Gamma(2\alpha+1))^3\Gamma(5\alpha+1)\Gamma(8\alpha+1)}$; $U_{1,\alpha}(0,9)=0$; $U_{1,\alpha}(0,10)=0$;

$U_{1,\alpha}(1,h)=0\ for\ h=1,2,3,\ldots$;

$U_{1,\alpha}(2,1)=0$; $U_{1,\alpha}(2,2)=0$; $U_{1,\alpha}(2,3)=-\frac{4c\Gamma(\alpha+1)}{\Gamma(3\alpha+1)}$; $U_{1,\alpha}(2,4)=0$; $U_{1,\alpha}(2,5)=0$;

$U_{1,\alpha}(2,6)=\frac{8c^2(\Gamma(\alpha+1))^2\Gamma(5\alpha+1)}{\Gamma(6\alpha+1)\Gamma(2\alpha+1)}\left[\frac{\Gamma(4\alpha+1)}{\Gamma(2\alpha+1)\Gamma(5\alpha+1)}+\frac{2}{\Gamma(3\alpha+1)}\right]$; $U_{1,\alpha}(2,7)=0$; $U_{1,\alpha}(2,8)=0$;

$$U_{1,\alpha}(2,9)=-\frac{32c^3(\Gamma(\alpha+1))^3\Gamma(8\alpha+1)}{\Gamma(9\alpha+1)(\Gamma(2\alpha+1))^2}\left[\frac{\Gamma(4\alpha+1)\Gamma(7\alpha+1)}{\Gamma(2\alpha+1)\Gamma(5\alpha+1)\Gamma(8\alpha+1)}+\frac{\Gamma(4\alpha+1)}{\Gamma(3\alpha+1)\Gamma(5\alpha+1)}+\frac{\Gamma(5\alpha+1)}{\Gamma(6\alpha+1)}\left\{\frac{\Gamma(4\alpha+1)}{\Gamma(2\alpha+1)\Gamma(5\alpha+1)}+\frac{2}{\Gamma(3\alpha+1)}\right\}\right]$$
; $U_{1,\alpha}(2,10)=0$

and so on

Now, from (3), we have

$$u(x,t)=\sum_{k=0}^{\infty}\sum_{h=0}^{\infty}U_{1,\alpha}(k,h)x^{1.k}t^{\alpha h} \quad (7)$$

Using the above values of $U_{1,\alpha}(k,h)$ in (7), the solution of (1) is obtained as

$$u(x,t) = x^2 + \frac{2\Gamma(\alpha+1)}{\Gamma(2\alpha+1)}t^{2\alpha} - \frac{4c(\Gamma(\alpha+1))^2\Gamma(4\alpha+1)}{(\Gamma(2\alpha+1))^2\Gamma(5\alpha+1)}t^{5\alpha} + \frac{16c^2(\Gamma(\alpha+1))^3\Gamma(4\alpha+1)\Gamma(7\alpha+1)}{(\Gamma(2\alpha+1))^3\Gamma(5\alpha+1)\Gamma(8\alpha+1)}t^{8\alpha}$$

$$-\frac{4c\Gamma(\alpha+1)}{\Gamma(3\alpha+1)}x^2t^{3\alpha} + \frac{8c^2(\Gamma(\alpha+1))^2\Gamma(5\alpha+1)}{\Gamma(6\alpha+1)\Gamma(2\alpha+1)}\left[\frac{\Gamma(4\alpha+1)}{\Gamma(2\alpha+1)\Gamma(5\alpha+1)} + \frac{2}{\Gamma(3\alpha+1)}\right]x^2t^{6\alpha} -$$

$$\frac{32c^3(\Gamma(\alpha+1))^3\Gamma(8\alpha+1)}{\Gamma(9\alpha+1)(\Gamma(2\alpha+1))^2}\left[\frac{\Gamma(4\alpha+1)\Gamma(7\alpha+1)}{\Gamma(2\alpha+1)\Gamma(5\alpha+1)\Gamma(8\alpha+1)} + \frac{\Gamma(4\alpha+1)}{\Gamma(3\alpha+1)\Gamma(5\alpha+1)} + \frac{\Gamma(5\alpha+1)}{\Gamma(6\alpha+1)}\left\{\frac{\Gamma(4\alpha+1)}{\Gamma(2\alpha+1)\Gamma(5\alpha+1)} +\right.\right.$$

$$\left.\left.\frac{2}{\Gamma(3\alpha+1)}\right\}\right]x^2t^{9\alpha} + \cdots \quad (8)$$

Take $c = 0$ in (1) and (8), the nonlinear inhomogeneous time fractional partial differential equation becomes linear and its solution is

$$u(x,t) = x^2 + \frac{2\Gamma(\alpha+1)}{\Gamma(2\alpha+1)}t^{2\alpha}, \quad (9)$$

which is same as obtained by Z. Odibat and S. Momani [18] using Generalized differential transform method.

4. Conclusion

This present analysis exhibits the applicability of the generalized differential transform method to solve non-linear inhomogeneous time-fractional partial differential equation. It may be concluded that GDTM is a reliable technique to handle linear and nonlinear fractional differential equations. This technique provides more realistic series solutions as compared with other approximate methods.

References

[1] Zhou, J.K.(1986).Differential Transformation and Its Applications for Electrical Circuits. Huazhong University Press,Wuhan, China.

[2] Chen, C.K.,and Ho,S.H.(1999).Solving partial differential equations by two dimensional differential transform method. *Appl. Math. Comput.*, 106:171–179.

[3] Ayaz,F.(2004).Solutions of the systems of differential equations by differential transform Method. *Appl. Math. Comput.*,147:547–567.

[4] Abazari,R.,and Borhanifar,A.(2010).Numerical study of the solution of the Burgers and coupled Burgers equations by a differential transformation method. *Comput. Math. Appl.*,59:2711–2722.

[5] Chen,C.K.(1999).Solving partial differantial equations by two dimensional differential transformation Method. *Appl. Math. Comput.*,106:171–179.

[6] Jang,M.J.,and Chen,C.K.(2001).Two-dimensional differential transformation method for partial differantial equations. *Appl. Math. Comput.*,121:261–270.

[7] Kangalgil,F., and Ayaz,F.(2009).Solitary wave solutions for the KDV and mKDV equations by differential transformation method.*Choas Solitons Fractals*,41:464–472.

[8] Cole, J.D.(1951). On a quasilinear parabolic equation occurring in aerodynamics. *Quart.Appl. Math.*, 9:225–236.

[9] Arikoglu,A., and Ozkol,I.(2006).Solution of difference equations by using differential transformation Method. *Appl. Math. Comput.*,174:1216–1228.

[10] Momani,S., Odibat,Z., and Hashim,I.(2008).Algorithms for nonlinear fractional partial differantial equations: A selection of numerical methods. *Topol. Method Nonlinear Anal.*,31:211–226.

[11] Arikoglu,A., and Ozkol,I.(2007).Solution of fractional differential equations by using differential transformation method.*Chaos Solitons Fractals*, 34:1473–1481.

[12] Soltanalizadeh,B.,and Zarebnia,M.(2011).Numerical analysis of the linear and nonlinear Kuramoto-Sivashinsky equation by using differential transformation method. *Inter. J. Appl. Math. Mechanics*,7(12):63–72.

[13] Tari,A.,Rahimi,M.Y.,Shahmoradb,S.,and Talati,F.(2009).Solving a class of two-dimensional linear and nonlinear Volterra integral equations by the differential transform method.*J. Comput. Appl. Math.*, 228:70–76.
[14] Nazari,D., and Shahmorad,S.(2010).Application of the fractional differential transform method to fractional-order integro-differential equations with nonlocal boundary conditions.*J. Comput. Appl. Math.*, 234:883–891.
[15] Borhanifar,A., and Abazari,R.(2011).Exact solutions for non-linear Schr.dinger equations by differential transformation method.*J. Appl. Math. Comput.*,35:37–51.
[16] Borhanifar,A., and Abazari,R.(2010).Numerical study of nonlinear Schr.dinger and coupled Schr. dinger equations by differential transformation method.*Optics Communications*,283:2026–2031.
[17] Momani,S.,Odibat,Z.,and Erturk,V.S.(2007).Generalized differential transform method for solving a space- and time-fractional diffusion-wave equation.*Physics Letters. A*,370(5-6):379–387.
[18] Odibat,Z., and Momani,S.(2008).A generalized differential transform method for linear partial differential equations of fractional order. *Applied Mathematics Letters*, 21(2):194–199.
[19] Odibat,Z.,Momani,S.,and Erturk,V.S.(2008).Generalized differential transform method:application to differential equations of fractional order. *Applied Mathematics and Computation*, 197(2):467–477.
[20] Ertiirka,V.S., and Momanib,S.(2010).On the generalized differential transform method: application to fractionalintegro-differential equations. *Studies in Nonlinear Sciences*,1(4):118-126.
[21] Garg,M.,Manohar,P., and Kalla,S.L.(2011).Generalized differential transform method to Space-time fractional telegraph equation. *Int.J.of Differential Equations,*Hindawi Publishing Corporation, vol.2011,article id.:548982,9 pages,doi.:10.1155/2011/548982.
[22] Bansal,M.K., and Jain,R.(2015).Application of generalized differential transform method to fractional order Riccati differential equation and numerical results. *Int. J.of Pure and Appl. Math.*,99(3):355-366.
[23] Das.S.(2008).Functional Fractional Calculus.Springer.
[24] Miller,K.S., and Ross,B.(1993).An Introduction to the Fractional Calculus and Fractional Diff. Equations. John Wiley and Son.
[25] Caputo,M.(1967).Linear models of dissipation whose q is almost frequency independent-ii. *Geophys J. R. Astron. Soc*,13:529-539.
[26] Podlubny,L.(1999).Fractional differential equations: An introduction to fractional derivatives, fractional differential equations, to methods of their solution and some of their applications. Academic Press.
[27] Almeida,R., and Torres,D.F.(2011).Necessary and sufficient conditions for the fractional calculus of variations with caputo derivatives.*Communications in Nonlinear Science and Numerical Simulation*, 16:1490-1500.

AUTHOR INDEX